CREATIVE LIGHT

PHILIPS TECHNICAL LIBRARY

CREATIVE LIGHT

L. C. KALFF

MACMILLAN

SBN 333 12339 5

First published in England by
THE MACMILLAN PRESS LTD
London and Basingstoke
Associated companies in New York, Toronto, Melbourne
Dublin, Johannesburg and Madras

PHILIPS

Trademarks of N.V. Philips' Gloeilampenfabrieken

No representation or warranty is given that the matter treated in this book is free from patent rights; nothing herein should be interpreted as granting, by implication or otherwise, a licence under any patent rights.

Printed in the Netherlands

PREFACE

Human life is unthinkable without light, which often dominates all our thoughts and activities, without our being aware of it.

The lighting industry has made remarkable progress during the past 70 years, and even today new light sources are still being developed and more efficient and refined methods of production applied. However, the author has found during forty years of activity in the science and art of lighting that the scientists and the technicians spent all their time perfecting light sources and the means of mass production, leaving it to a few individuals to learn how best to apply these light sources in the service of man.

In the 19th century, light was not very much appreciated by the public (except perhaps by the farmer) and it was not yet considered to be an important factor in creating our mental and physical health. In those days, windows in buildings and houses were taxed, as if they were a luxury. Ladies protected their complexion against sunlight with parasols and face powder. Daylight in the houses was filtered through two or three layers of curtains. And workers in factories were even worse off: only with the gradual development of social reforms was the right to better living and working conditions for everybody granted. Even at the beginning of this century, the minimum requirements for the lighting level during work were absolutely inadequate. It was only after World War I that people started to enjoy sunshine and that a sunburnt face was no longer a disgrace. It seems almost unbelievable that this was only 50 years ago.

Parallel with this growing desire for light, in the daytime as well as at night, lighting techniques developed but it gives food for thought that the greatest stimulant to this development was the bitter necessity of the Second World War.

Too much attention is still being paid to the technical and economic aspects of lighting, and insufficient interest is taken in the well-being of man as the first aim of technical developments. We therefore hope that this book will be read not only by architects, decorators and lighting experts, but also by electrical engineers.

We should never forget that all technical developments meant to increase the comfort of our fellow men, such as ventilation, heating, air conditioning, acoustic engineering,

the fight against water and air pollution and lighting, do not exist in the first place to fight discomfort but *to contribute to human well-being* in the positive sense!
There are too few specialists in the lighting field throughout the world who are occupied with better lighting in this sense.
However, these few have always sought contacts with one another so as to make an international team which profits as much as possible from each others' thoughts and work. I would like to take this opportunity of extending my hearty thanks to this group of perhaps a few score of men for many years of co-operation.
I sincerely hope that this modest contribution to the art of lighting may bring others to pay more attention to this modern technique with its many possibilities of new applications.

L. C. Kalff

October 28, 1970

CONTENTS

INTRODUCTION

a. Light is emitted from a light source towards an object.

b. That object reflects the light, a part of which travels in the direction of the eyes.

c. That part of the reflected light which enters the eye is focussed by the lens of the eye to form an image of the object on the retina.

d. The light impinging on the retina is absorbed by the receptors — the cones and the rods.

e. The light thus absorbed is converted into pulses, which are transmitted to the brain via the optic nerve.

f. In the brain these pulses are compared with previously received visual information and are then dismissed as unimportant, or used to motivate action, or stored as memories.

Light, the all important element in this process, is characterized by:
- colour (spectral distribution)
- intensity (amount of flux per solid angle)
- direction (propagation in straight lines)

The object is characterized by its shape, colour and texture, and whether it is opaque or translucent. When light impinges upon it, this light is absorbed, reflected or filtered, thereby producing changes in colour, intensity and direction.
This is a *physical* process.

The absorption of the light at the retina, its conversion into pulses and their transmission to the brain are mainly *physiological* processes. The conversion of pulses into impressions and the interpretation of these impressions are of a *psychological* nature. This book deals mainly with light, but also with vision and the mental interpretation of what is seen.

I The first chapter shows how buildings, through the ages, demonstrate the tremendous influence of the light (climate) on their character (shapes, silhouettes, colours, surfaces, details).

II The second chapter indicates that in the interiors of these buildings the light can also be used to create impressions of space and atmosphere. This has been recognised at various periods and by a few architects. It can be considered as the highest expression in architecture — real "light architecture".

III With the development of modern construction techniques and also due to economic considerations, architects are now more restricted as to the possibilities of creating "light architecture". With the introduction of the industrial era other requirements have come to play an important role.

IV No rules or indications can be given for good visual conditions without some knowledge of the functioning of the eyes.

V The design of the visual field, based upon this knowledge, is explained in this chapter. The function and construction of the retina has a close connection with the composition of the visual field in terms of line, colours and brightness.

VI The brain, the mind, plays an important role in vision. We are, for instance, greatly influenced by the diurnal cycle of day and night, of work and recreation, each activity and hour demanding a special visual environment.

VII The quantities and qualities of the light we need or prefer during our work or other activities of our daily life are difficult to define. Since perception is a mental process it is impossible to determine the ideal lighting from laboratory experiments. Threshold values and absence of discomfort never guarantee comfortable vision.

VIII With the improvement of lamp-making technique, and the economic possibilities of using considerable amounts of artificial light, we now possess the means to bring about almost any visual condition by day and by night; but we are still only beginning to master the art of providing the ideal visual conditions for our work.

IX It is not only our eyes which tell us what and how we like to see. Our mind has a great influence on our visual perception, which means that the whole human being, with its experience, its wishes, interests and aversions, influences the way we see.

About the ways in which we can study these problems so that gradually we shall be able to design good and comfortable visual environments.

X Technical improvement of light sources and lighting equipment will go on, but it will never lead to perfection. It is only by studying the reactions of the users of installations that we will be able to come nearer to perfection. This needs human interest, research and experience.

XI Epilogue.

Fig. 1. In architecture a dry and sunny climate leads to simple and big shapes. Egyptian temples, built all in granite without any variation in colour, have sufficient character by their shape alone. (Edfu)

Chapter I

CLIMATE AND SHAPES IN ARCHITECTURE

When we look at a building we realize that its aspect changes greatly with the variations of daylight. It may be dull under clouds and in the rain: colours and details may bring out its delicate and cheerful beauty in sunshine.

If we look at buildings in different climates and latitudes we must come to the conclusion that architecture is greatly influenced by the climate.

Where sunshine and a clear sky are the prevailing conditions it is obvious that architects conceive their buildings under these circumstances, and in countries where mist and clouds, humidity and a sun that does not rise high in the sky cause a much less strong, more diffuse and cool light, the buildings are of different shapes, colours and textures.

The merciless African sun certainly caused the Egyptian architects to choose massive and utterly simple shapes for their pyramids, temples and palaces (Fig. 1).

Enormous flat walls and huge round columns are their main features, whilst simple cornices produce one straight and heavy shadow to outline the silhouette of these buildings against the sky.

On these huge granite surfaces we find only the very thin and finely engraved lines of the hieroglyphs, which would be invisible in the absence of harsh Egyptian sunshine (Fig. 2).

Around the Mediterranean we also find Greek and Roman architecture.

The Greeks, in their dry and sunny climate, first created the highly refined marble shapes of columns: slender and strong, with delicate fluting, the aspect of which continuously changes with the changing orientation of the sun. They also employed intricate cornices — a composition of mouldings, blocks and sculpture, visible in every detail by light, shadow and half shadow, and also changing every hour of the day with the rising and setting of the sun (Fig. 3).

These architectural details have played their parts through 20 centuries all over the world and it is clear that they have been applied in climates and countries quite

Fig. 2. On these temples the inscriptions and reliefs are very shallow and do not have any influence on the grandeur of the shapes. In the absence of the harsh sunlight the inscriptions would hardly be visible. (Dendera)

different from those from which they originally came.
Roman architecture spoke very much the same language, although construction with arches and vaulting and the different stones used instead of the Greek marble, already has an influence on the use of mouldings and columns. Characteristic for both periods are the fine details, the delicate stone carving and the uniformity in material and colour, since the sunlight was sufficient to show to advantage the whole design (Figs. 4 and 5).

Later the same shapes and architectural details were used profusely in the Renaissance, which started in Italy, taking the Roman architecture as an example. As long as this style stayed in Italy everything was all right, since the shapes and compositions were used in the same climate as the originals (Fig. 6). But in the 16th and 17th centuries the Renaissance crept gradually northwards and the fashions in France, Belgium, Holland and Britain started with the same elements but under a completely different sky, with more mist, clouds and less sunshine.
What had an excellent effect in the Mediterranean atmosphere was much less expressive in the more northern countries. Architects therefore had to look for new and different means of expression, and found them in the use of colour, different building materials and textures, higher reliefs and deeper mouldings (Fig. 7, 8 and 8A).
They also paid much attention to the silhouettes of their constructions which are always clearly visible against a cloudy sky (Fig. 9).
The most charming examples of the "adapted" Renaissance can be seen in Flanders, Holland and Denmark. Here we find the red brick facades with white limestone ornamentation and gables, copperclad domes and roofs with their beautiful green patina, and the wonderful towers with their filigree silhouettes in stone and lead covered wooden lacework (Fig. 10).
If we compare the famous Amsterdam towers with the Campanile on the Piazza San Marco in Venice, the difference is obvious: in Amsterdam the most elaborate silhouettes, in Venice the simple and austere geometric form, almost without any adornments (Figs. 11, 12 and 13).

In Mediaeval architecture we find the same differences between buildings of the period in Italy and Spain and those in northern countries, between 45° and 65° north latitude.

The Gothic churches and buildings in the north show in all their details how very great has been the influence of the climate on their architecture. Towers, buttresses, gables all with finials, gargoyles and filigree balustrades make a very lively silhouette against the clouded sky. Very deep cornices and mouldings make lines of heavy shadow notwithstanding the diffuse light (Figs. 14 and 15). The hundreds of statues of saints in the deep portals of the cathedrals have similar deep carvings that make them clearly visible even in the shadow (Fig. 16). We here think in the first place of the cathedrals in Paris, Chartres, Beauvais, but also those of Antwerp, Bois le Duc, London, Cologne, Ulm (Fig. 17).
We find very similar shapes in the duomo of Milan, but we consider them crude and harsh in the bright sunlight of Italy.

3

Although this chapter deals only with the influence of daylight on the exterior of buildings, it is nevertheless necessary to mention the windows, since they form an important feature in the design of the buildings (Fig. 18).

In Greece, North Africa, the south of Spain and Italy, the windows are often deep and small openings in the thick walls that have to protect against the sunshine and heat (Fig. 19).

The further we travel northward the bigger are the windows and the thinner the walls. This is already noticeable in the moist atmosphere of Venice, but the 17th and 18th century houses in Amsterdam are a very true expression of the climate in Holland. They admit a maximum of daylight, and the heavy white-painted wooden frames, between the narrow red or brown brick piers, are the most prominent features in this architecture (Fig. 20).

Similar windows can be found in the Georgian houses in England and Ireland; also in Copenhagen and in Prague: all places with similar climates, with much mist and clouds, and long winters (Fig. 21).

Fig. 3. The Greek doric temples have an ever-changing aspect to the changing orientation of the sun during the day. (Athens)

Fig. 4. All Greek temples have simple and prominent details; the fluting of the columns, the moulding of the cornice, the triglyphs in the frieze, are all clearly outlined in the direct light of the sun which, by its movement, continuously changes the appearance.

Fig. 5. The "Casa de las Conchas" is a very simple and massive block. Its whole character is due to the pattern of shells and their strong shadows on the walls. (Salamanca)

Fig. 6. San Firenze, Florence.

8

Fig. 7. Renaissance architecture in the less sunny countries uses lively silhouettes against the grey sky, vivid colours of red brick and white stone, and grey or green roofs clad with lead or copper, in order to obtain striking effects notwithstanding the absence of sunshine and shadows. (Haarlem)
Fig. 8. Moreton Old Hall (Cheshire).

8A
9

8B

Fig. 8A, 8B. Zaanse Schans (Zaandam).
Fig. 9. The Exchange in Copenhagen shows how the Renaissance architects used colour and silhouettes under a clouded sky in order to obtain an expression which the architects in Italy and Spain could realize by sunshine and shadow.

Fig. 10. The big windows, with their heavy white frames stand out clearly in the dark-red or brown brick walls. The elegant gables are seen as dark silhouettes against a grey overcast sky. The Renaissance architecture in the northern countries is completely different from what we find around the Mediterranean. (Amsterdam)

11

12

Fig. 11. The famous church spires of Amsterdam with their vivid colours and their lively silhouettes are always gay elements in the skyline, even when it rains. (Amsterdam)
Fig. 12. The Italian towers are usually simple and austere (Venice).

Fig. 13. The dry and sunny climate in Spain also lends itself to simple outlines and little decoration or colours. (Escorial)

14

15

Fig. 14. The Gothic cathedrals in the north of France, Holland, Germany and England are rarely seen in brilliant sunshine. They are beautiful under a clouded sky. The silhouettes of all these details are typical for these climates. ('s-Hertogenbosch)

Fig. 15. The very deep mouldings and the filigree balustrade have been created for the misty atmosphere of this latitude. ('s-Hertogenbosch)

17

Fig. 16. The "Portail Royal" of the Chartres Cathedral is almost permanently in shadow, and the climate in this part of France is not always sunny. The statues and the sculpture have been perfectly adapted to the prevailing conditions; the shapes are well defined, the moulding and the relief are deep, so that even in intense shadow the figures do not lose anything of their expression.
Fig. 17. Grencester Church of St. John Baptist.

Fig. 20. The fronts of these 18th century houses are very functional. They have a minimum area of brick walls and a maximum area of windows that admit the soft and filtered light of Amsterdam.

Fig. 21. The facades of the 17th and 18th century houses in the Beguinage in Amsterdam clearly show that the inhabitants were in need of a maximum amount of daylight.

Fig. 22. Simple shapes and smooth surfaces of one colour become visible in the light admitted by hardly visible windows. They create an atmosphere of austerity and meditation. (Fontenay)

Chapter II

LIGHT, THE CREATOR OF SPACE AND ATMOSPHERE

> "Light and shadow are the loudspeakers of this architecture of truthfulness, calmness and strength"
>
> Le Corbusier

As shown in the first chapter, the specific qualities of the light have a great influence on the external shapes in architecture, that is to say the architect's design is influenced by the local climate.

Interiors show similar influences, but here the light mainly creates impressions of space and atmosphere, rather than forms and volumes. Successful use of daylight in interiors is the exception rather than the rule (Fig. 22).

We will now mention a few periods and buildings that have made the strongest impression on us.

The first example is the architecture of the Cistercian order in the 12th century. The disciples of St Bernard then started to build monasteries and chapels in many places in Europe. The principles of the order, namely utter simplicity, sobriety and purity, are clearly reflected in the way of building, especially during the first centuries after the foundation of the order (1098 in Citeaux). The heavy, flat, stone walls exclude all influences from the world outside. Massive stone columns support simple arches and barrel vaults; no decoration or sculpture is employed (Fig. 23).

The windows are placed high in the thick walls, so that they make deep niches which screen off the bright sky on either side so that the spectator in the central nave does not see the windows or the sky. The light from outside is first reflected in the niches before entering the church, and is soft and diffuse (Fig. 24).

When entering these buildings, be it a chapel, a refectory or a dormitory, one is impressed by the simple clearness of the architecture in which a beautiful light, coming from invisible windows, touches in delicate gradations of brightness the walls, arches and columns that enclose the space. The buildings everywhere show the same rough light warm grey stone. All the rooms are filled with a fine light, showing every surface with well proportioned and functional brightness (Fig. 25).

The parts that have to express strength, such as the columns and arches, are dark; the floor, the choir, niches and side chapels are bathed in a subdued and serene light. It is all clear and logical and simple, yet conveys no impression of being deliberately designed to please.

Fig. 23. The Cistercian buildings of the 12th century can be found all over Europe. They all have the same atmosphere of devoutness. (Heiligenkreuz near Vienna)

Fig. 24. Even in the dormitories of this Cistercian monastery the light entering the department is soft, notwithstanding the fierce sun of the Provence. It is softened by the multiple reflexions in the deep niches of the windows. (Le Thoronet)

Fig. 25. Twelfth-century Cistercian buildings are also in Spain. There again the light lends a rich expression to the simple and modest interiors. (Poblet near Taragona)

24

25

26 27

Fig. 26. The Church of St Nicholas in baroque style shows how the architect can obtain striking and theatrical effects, if he has learned of all the possibilities of light. (Prague)

Fig. 27. The architect has managed to create an impressive succession of spaces by means of light, coming from invisible windows. They light up the central dome and the choir so that the altar with all the statues around are silhouetted against a bright background. (Church of St George in the Benedictine Monastery of Weltenburg)

Since this order of monks had settlements in so many countries one often is surprised to find this "light architecture" in unexpected and remote places. One finds it in the first place in France: Fontenay near Langeais, le Thoronet near Toulon; in Poblet near Tarragona in Spain; in Austria in Stift Heiligenkreuz. The examples mentioned here are all dated from the first years of the order. Later the monks had more contact with the outer world and adopted styles that rendered this utter modesty and simplicity impossible (Gothic, Renaissance, Baroque).
In the 12th century the Cistercians were certainly masters of light architecture. It is curious to see how this art disappears from time to time and then for no apparent reason pops up again in a completely different place and period.
The Baroque, of which one finds so many examples in the south of Germany, Austria, Czechoslovakia and Switzerland, shows a completely different attitude from the Romanic architecture in France. It exhibits pomp and splendour and richness, but also has a tendency towards the excessively ornate at times descending almost to the tawdry. The architect appears to have been out for a kind of theatrical ecstasy which would have been completely unacceptable to the first Cistercians.
Some masters in baroque architecture, however, demonstrate that it is possible to obtain the most dramatic and theatrical effects of *clair obscur* and spatial expression by a judicious use of the light. They could obtain at will light and shade, the required direction and quantity of light, the effects of form and colour by choosing the proper place and dimensions for the windows.
As an example of this period in which the light played such an important part we may mention the St Thomas Church in Prague. The site is in itself not very good, since the church abuts on one side against a monastery so that the symmetrical interior receives light from one side only, and one aisle is much darker than the other.
Nevertheless the visitor obtains a striking impression of vastness and space when entering the building (Fig. 26). The nave is very high, the walls end in a mighty cornice at about 22 metres from the floor, over which is seen the vault of the roof. The cornice is more than a metre wide, and it frames the ceiling which is light in colour, and decorated with celestial scenes, with angels and saints floating between

clouds in the sky. This vault is lighted by semicircular windows over the cornice, these windows being invisible from below. This effect of a bright and weightless ceiling is emphasised by the contrast with the heavy mouldings of the cornice which casts dark shadows.

The walls of the nave rest on heavy columns which form dark volumes between which one sees the light, coming from the deep windows in the aisle.

Those windows light the vaulting in the aisle, the niches in which the windows have been placed, and thus, indirectly, the outer wall of the aisle itself.

As already stated, the left aisle is without windows. The outer wall of this aisle therefore has to receive its light from the opposite windows. This wall, like the main vault, has been decorated with celestial scenes which, with their light colours, obliterate the limitation caused by the wall.

In this way the architect has achieved a dramatic and exalting effect by a masterful combination of light, volumes and colours.

When looking at the altar from the centre of the church one gets an impression of standing in a grove of heavy beeches; looking up one sees the sky between the branches, and between the dark treetrunks one glimpses the luminous distance.

This we believe to be light architecture par excellence, showing a mastery of the art equalling that of the 12th century Romanic architecture, though of course with a completely different aim. The baroque architect wishes to create a certain effect and uses for that purpose all the means at his disposal; the Cistercians simply created a sacred space in stone and light (Fig. 27).

Other fine examples of light architecture can be found in the domestic architecture of the 18th century houses on the canals in Amsterdam and other Dutch towns.

The art of making larger glass panels rendered it possible to provide bigger windows — a very desirable feature in this rather sombre climate.

The sites are generally narrow: 5 to 8 metres wide; the rooms, especially on the drawing-room floor, are deep and high, so the windows had to be as high and wide as possible. Those windows, in contrast with the church windows, had to allow for a free contact with the outside.

When looking out of these windows, there is a great contrast between the wall and the sky. This contrast is, however, successfully softened by mouldings in the window frames and in the glazing bars (see Figs. 28, 28A and 28B).
Because of the soft soil in Amsterdam, the walls of these houses are thin and light, but the windows are often placed in deep niches formed by the folding shutters, which, painted white, provide by day a very fine intermediate tone between walls and sky (Figs. 29A and 29B). The sun can penetrate only when it shines straight on to the house. In summer the big trees along the canals filter the direct sunlight and admit a cool greenish light to the rooms. This kind of window is also found in many other towns in England, Ireland and France.
When investigating Amsterdam domestic architecture more closely, light effects are found which remind us that we are in the country of Vermeer. The long corridor leading into the house is generally clad with white marble; doors are often richly carved, and in dark wood or colour stand out against the white walls. At the back of the houses there is often a garden room, and over it the drawing-room. In that case the corridor runs along a small inner court to these rooms, receiving a cool light through a big window looking out upon that court. Seen from the front door, this light gives a charming and beautiful accent in the long corridor. Similar effects of light and space we find in many staircases in England, where hidden windows on the landings create charming accents of light at the end of the corridor (Fig. 30).
The art of using daylight as an architectural element creating space and atmosphere is alive up to the present time. We certainly find it in the work of the late French architect Le Corbusier. He saw the spaces he was creating together with the light that was to be part of it. The Ministère du Travail in Rio de Janeiro, which he conceived in 1939, clearly shows that he considered the way the daylight would enter the building one of the principal motifs in his architecture (Fig. 31). The hot climate causes the sun to heat up the enormous facades to an unbearable degree. It would have been impossible to build this very high building with heavy and heat absorbing walls. So he invented the "Brises soleil" — deep baffles which screen off the sunlight so that the walls and the windows are in shadow.

28

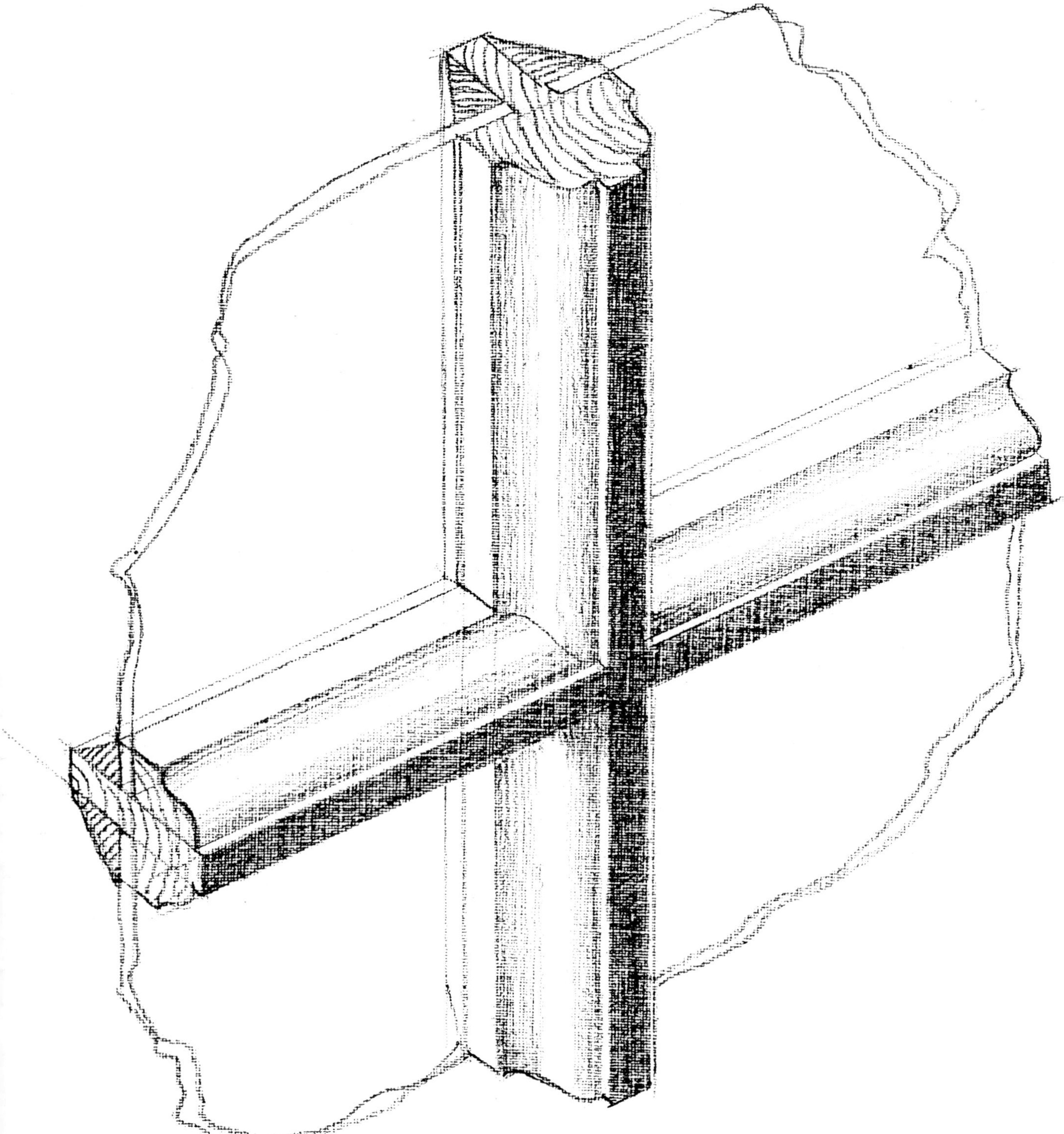

28A

Fig. 28. In this 18th century drawing room the windows, with their moulded glazing bars and frames and the deep jambs made by the folded shutters, look out on the canal and the trees and houses opposite. The light painted shutters soften the heavy contrast between the dark walls and the bright sky. (Amsterdam)

Fig. 28A. The glazing bars in the 18th century windows were made of wood, often painted white. The shadows and half shadows in the mouldings of these bars reduce the strong contrast between the bright sky and the dark surface of the bars facing the interior.

Fig. 28B. Modern steel or bronze window frames and glazing bars have a rectangular section, so that there is always a harsh contrast between the dark metal and the bright sky.

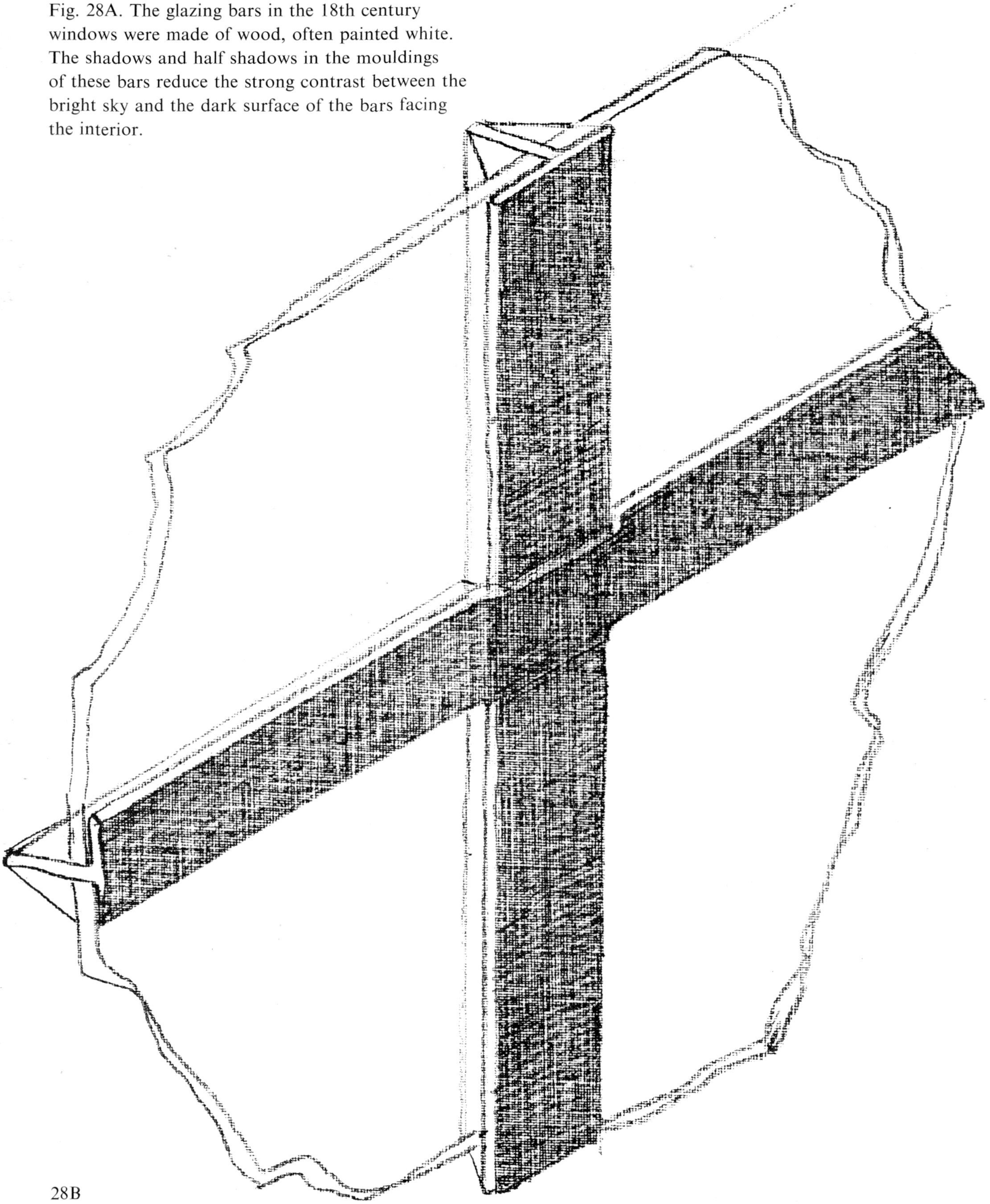

28B

29

Fig. 29. This room has only two windows; we see that the dark pier and its shadow cut the room in two. We find here once more the beautiful windows mentioned in Fig. 28. (Amsterdam)

Fig. 29A. The brighter parts of a room always attract our attention more than the darker areas. A room with three windows always has a centre of attention, which is accentuated by a maximum of light.

Fig. 29B. If a room of a similar size has only two windows, the central place remains relatively dark while there are two well-lighted zones in front of the windows. This accentuates the end walls so that the space is divided into equivalent parts (see Figs. 28 and 29).

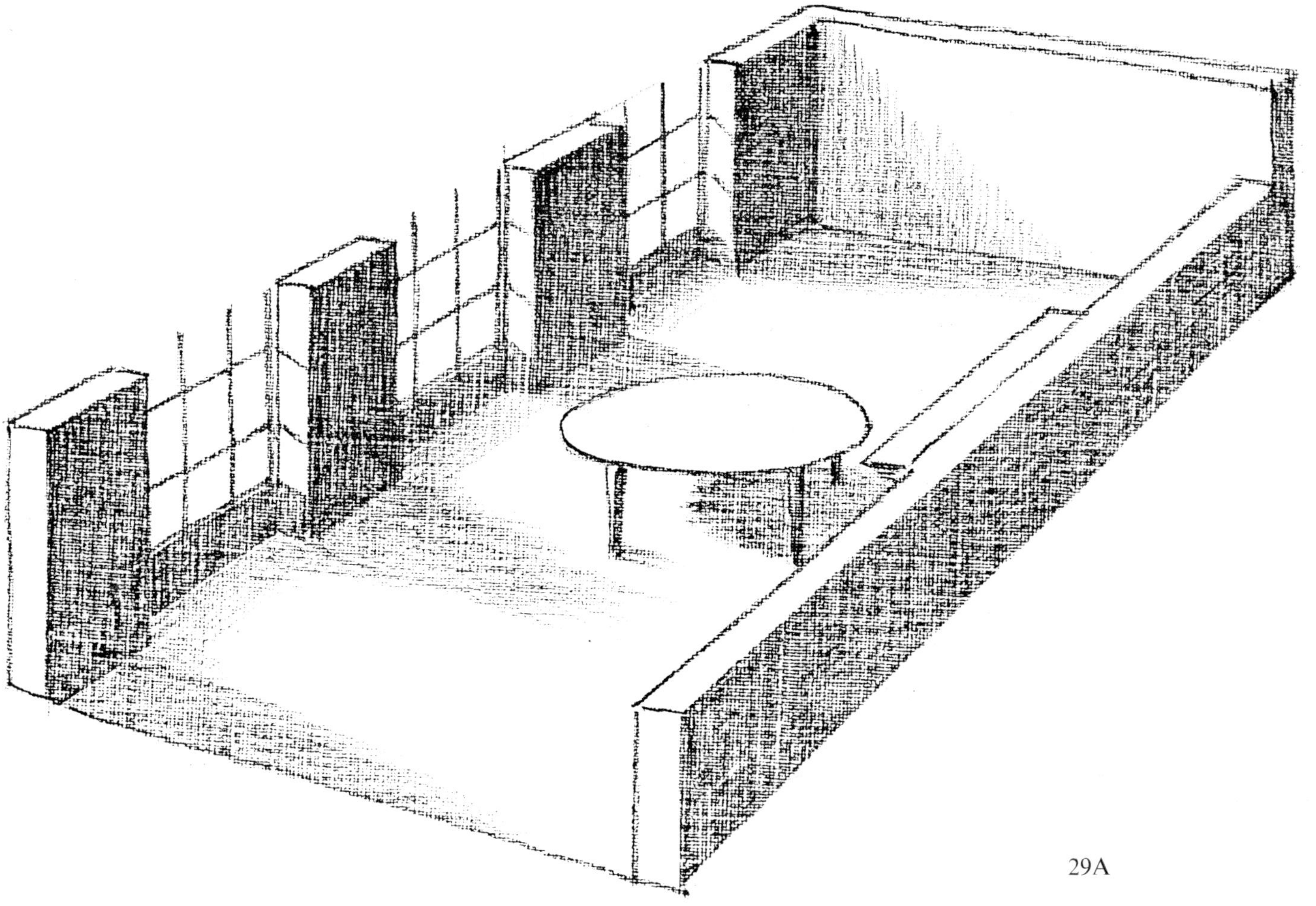

29A

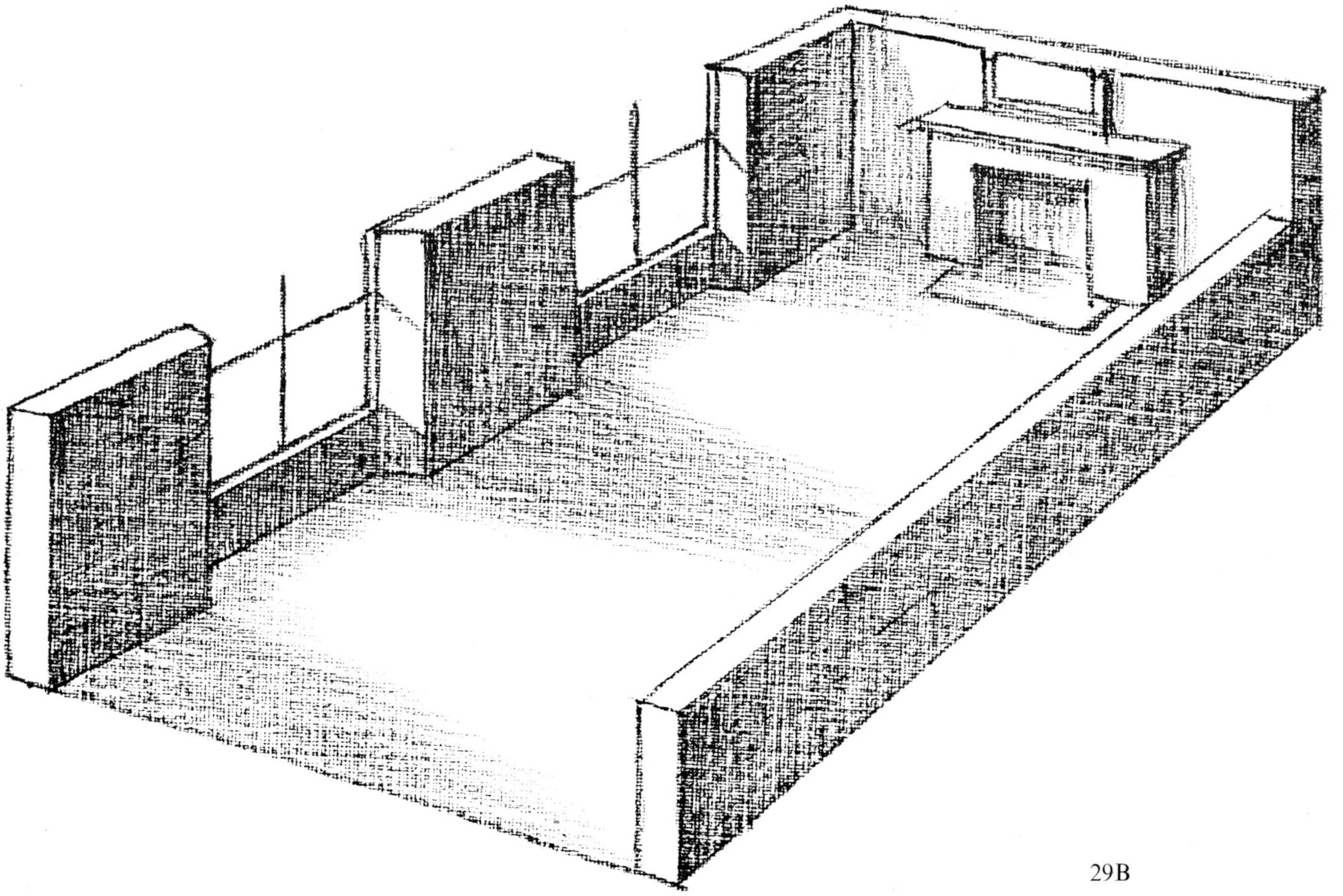

29B

Fig. 30. Somerset. Bath, Circus No. 10.

Fig. 31. In 1937 Le Corbusier invented, for the Ministry of Works in Rio de Janeiro, the "Brises soleil", a facade of screens in front of th walls of the buildings, in order to protect it agai the hot sun. Since then this feature is common in the South American and Mexican towns.

The diffuse light, coming from the sky, can enter freely in an almost horizontal direction, producing a very unique light effect in the interiors.
As we have said, Le Corbusier always designed his interiors together with their lighting as an integrated whole. We had the privilege of discussing lighting problems with him on many occasions.
He then took a yellow and a violet pencil — yellow for light, violet for shadow — and with these he indicated on plans and sections the parts he meant to be light or dark (Fig. 32).
The chapel of Ronchamp, the La Tourette monastery near Evreux, the government buildings of Chandigar, were all studied this way, after which the arrangements for daylight and artificial light were designed (Figs. 33 and 34).
These constructions in their turn became motifs for creating new shapes, such as the semicircular light tower in the chapel of Ronchamp and the "light cannons" in the La Tourette monastery (Figs. 35 and 36). He greatly objected to defacing the surface of the concrete ("ne détruisez pas la noblesse de mon beton!") for installing electric wiring, because he knew that the "making good" will always show afterwards. Often he wanted to have artificial light entering the rooms in the same way as daylight. In order to achieve this, the light sources had to be placed in niches and windows so that the brightness pattern of his original conception would be the same at night as by day.
We should like to end this little survey of examples of "light architecture" with one beautiful example we found in Helsingör, Denmark. This beautiful Renaissance castle, the Kronborg, built in 1635 by the Dutch architect Antoon van Obbergen, lies along the Sont, surrounded by a vast open stretch of land and water. It is built in that typical Dutch Renaissance style, using red brick and white stone which is also found in Copenhagen and Amsterdam.
On the first floor there is a magnificent knight's hall: long, with white-washed walls, tall windows arranged symmetrically on both sides in the very thick walls. The windows look out over the wide country, but looking along the length of the hall, the deep white niches screen off the direct light from outside. The abundant diffuse light creates an unforgettable atmosphere in this well-proportioned stately room (Fig. 37).

Fig. 32. The light in the Chapel of Ronchamp enters through windows in a very thick wall. The light is coloured by stained glass and makes a wonderful pattern of varied brightnesses and colour in the deep niches of the windows. The curved roof loses its "heaviness" by reason of a narrow strip of light that separates it from the walls.

Fig. 33. The semi-cylindrical towers and belfry of the chapel of Ronchamp became now characteristic features of the architecture of Le Corbusier.
Fig. 34. For the chapel in Ronchamp Le Corbusier created semi-cylindrical towers. Through them the light penetrates the chapel and illuminates the wall behind the altar.

35

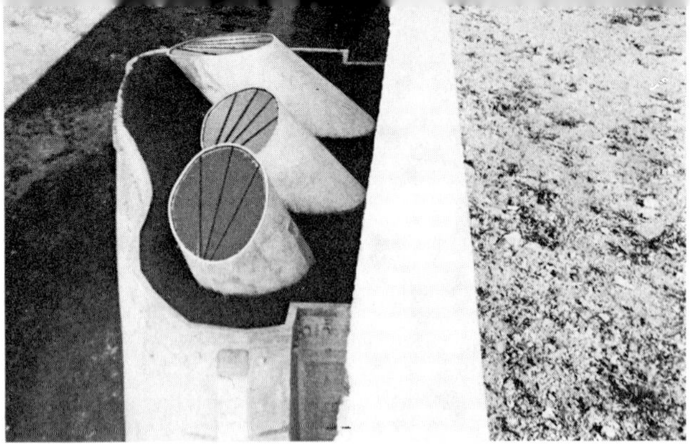

36

We now come to other more dynamic effects that have been obtained with the help of light.

In this connection we would mention a theory published in 1918 in a book called "Plastik und Raum" by Prof. Brinckmann.

He says that the interiors of monumental buildings produce their main effect by the successive impressions the spectator receives when entering and going through the different spaces.

For instance, one might enter the building through a high and spacious hall and then, continuing through a lower and narrower corridor, come into a wider space, lying square to the corridor at the end of which the eye can wander upwards in a high round central hall, covered with a dome and widened by niches that break up the walls.

In order to be able to analyse this succession of impressions he suggested a new idea, the "Raumkurven" (space curves) — lines that one can draw in the plan as well as in the sections of the building, indicating the curves our eyes (and the mind with them) follow whilst traversing the building. We have always found that these "Raumkurven" can help us considerably when analysing and appreciating important monumental interiors (Fig. 38).

We can for instance, describe the dynamic effects the architect obtains and which we experience when walking through a succession of different spaces. We come from outside into a relatively small entrance hall where we accommodate to relative darkness; we then come to a bright and wider space after which we are again enclosed by a narrow and longish corridor; and finally we experience a feeling of liberation in the high and wide central hall.

It is certain that this is what the architect had in mind when he made his plans. We wish to point out that this succession of impressions will have a far greater effect if the light and the colour in the different spaces follow the trend of the "Raumkurven": when widening, more light; when getting narrower or lower, less light; so giving greater emphasis to the spatial composition. Here again this is a refinement in the use of light in buildings which we may call "light architecture".

Fig. 35. Here we see the "light cannons" of Fig. 36 from the outside.
Fig. 36. For the monastery of La Tourette, Le Corbusier invented "light cannons" through which he could direct light on to the altars in the chapel. (Evreux)

37

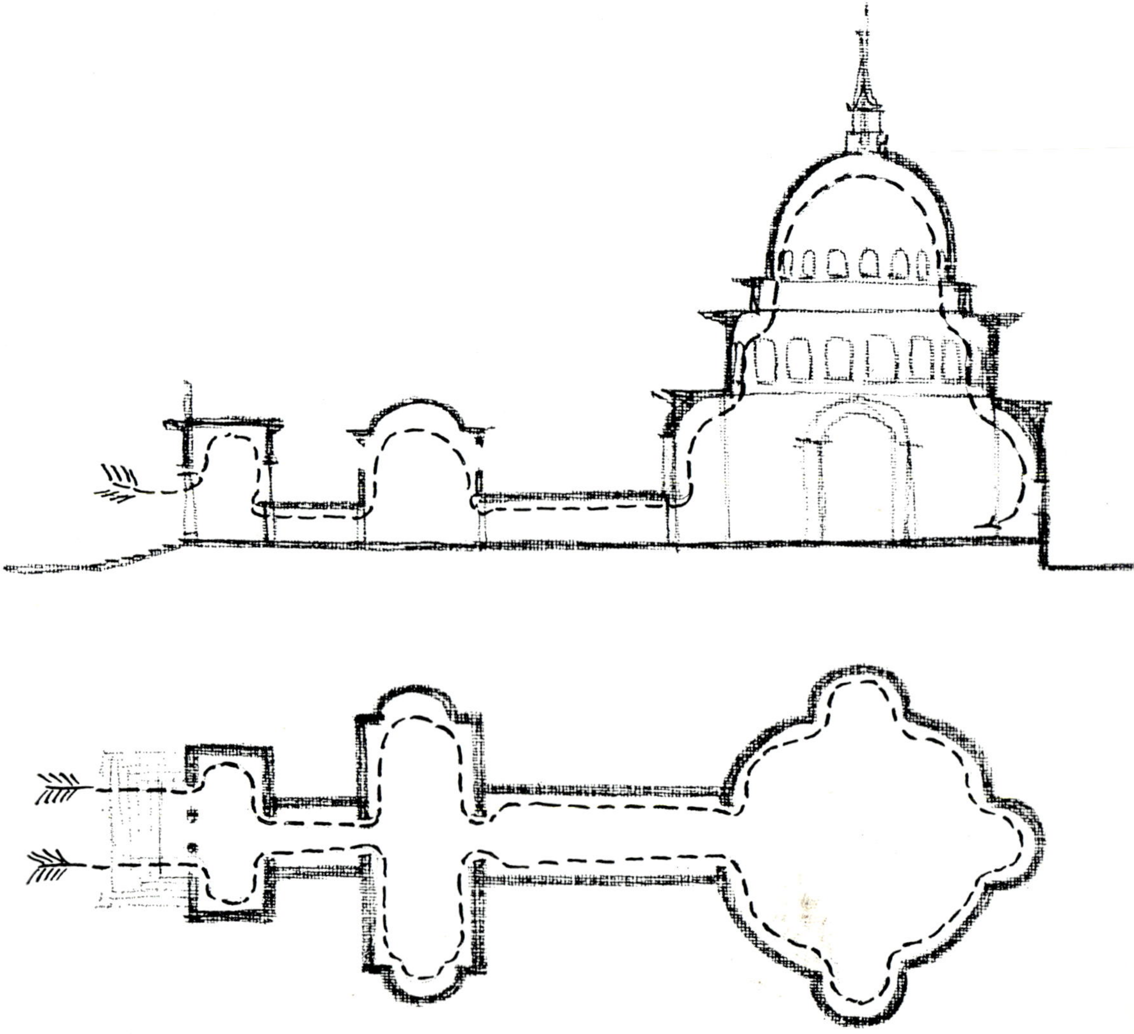

38

This term was invented in about 1920 by Professor Teichmüller of the Lichttechnisches Institut in Stuttgart. The word is a suggestive one, and it soon came into general use, but in our opinion, not in the right sense.

In those days the incandescent lamps were often hidden behind diffusing refracting glass panels. The interior decorator used these panels to build hollow glass volumes, translucent and lighted from within.

Girders, columns, walls and ceilings were treated in this way — a mock "light architecture". We do not want the reader to confuse this with what we consider the true meaning of the term as previously explained.

About 1925, after the Exposition des Arts Décoratifs in Paris this kind of luminous decoration was for a short time very popular. Highly favoured in those days were the pressed glass ornaments by the French manufacturers Lalique and Sabino (Fig. 39).

In interiors, solid volumes and shapes seldom have a similar effect to that which they would have in full daylight, since the light coming from windows is so much weaker and much more diffuse. That explains why in interiors we only rarely see shapes that have been created primarily to take advantage of well defined illumination.

One striking example, however, we do know. The windows in late Gothic churches become higher and higher, but they remain, for reasons of construction, rather narrow. That means that the daylight is allowed to enter through vertical strips of window. The clustered columns in late Gothic churches consist of cylindrical shafts, thus showing a strong vertical profile. The narrow vertical windows produce strong highlights and shadows on these shafts, which could never be so strong with windows of different dimensions (Fig. 40).

As a result of this lighting the columns become solid and independant objects, breaking up the space in the church by their large volumes. If we compare these

Fig. 37. Castle of Helsingör, Denmark.
Fig. 38. In this sketch of the plan and the section of a monumental building have been drawn the "space curves" (Raumkurven) which indicate how the observer, when traversing the successive spaces, receives alternate impressions of oppression and of relaxation, culminating in the free openness of the dome. It is felt that these curves also suggest the amounts of light and the brightnesses that should accompany and accentuate the succession of the different spaces.

Fig. 39. A rather false "luminous architecture". About 1925 it became the fashion to build hollow pillars, girders and walls from pressed glass elements. By lighting up these elements from behind with glowlamps one could produce a form of luminous decoration which, however, was quite foreign to the real architecture of the building.

Fig. 40. The columns in Gothic churches are often bundles of small columns. They are no longer part of a wall (see Fig. 24) but they become individual volumes in the interior space of the churches. Their complicated shapes are accentuated by the light coming from the narrow and high Gothic windows, which produces a very striking clear obscure. (Bourges)

Fig. 41. The classical columns were created for sunlight. In interiors all these intricate details become hardly visible because of the soft and diffuse light; the shapes of the columns therefore become rather functionless.

Fig. 42. In interiors architects have often tried to give more expression to these columns by replacing the effect of the shadows by shiny materials and different colours; the shafts in polished marble, the bases and the capitals in bronze or gold. (Library in the Benedictine Monastery of Wiblingen)

columns with those which we find in the Cistercian chapels the difference is striking, because there the columns have a simple square section and they are really part of a wall in which openings give access to the side aisle.
There is another use of shapes in the interior we wish to mention here.
Greek and Roman buildings with their fluted columns and horizontally moulded cornices have been created obviously for the clear and sunny climate of the Mediterranian. The course of the sun around the fluted columns gives each hour of the day a different aspect to the fronts of the temples. The cornices with their horizontal mouldings have their own rhythm of light and shadows that also varies with the position of the sun.
Now in the Renaissance period and later also in the Classical era these architectural motifs, originally intended to be seen in full sunlight, are used for decoration in interiors. Of course, the light coming from windows has directions and quantities completely different from exterior light and therefore these shapes become more or less meaningless (Fig. 41).
As the details of these columns and cornices can hardly be seen in the diffuse half shadow under a ceiling, architects have sometimes used different materials and colours to give more meaning to the different elements. For instance, we may find polished red marble shafts with bronze capitals and bases, and polychrome cornices with gilt ovolos (Fig. 42).
Sometimes this seemingly meaningless application of architectural shapes in interiors does have a purpose.
A century or more ago, evening feasts were illuminated with oil lamps and candles. The number of candles, and with it, the amount of light was limited because the candles gave off a great amount of heat. A festival hall like La Galerie des Glaces in Versailles could be illuminated only by a few hundred candles. The architect had therefore to make the most of these by using huge crystal chandeliers, mirrors and gilt mouldings for sparkle, and white walls for reflection (Fig. 43). The guests in this hall also played their part in such a feast by wearing light coloured silks and lustrous satin.

It is interesting to reflect that there are interiors that receive no daylight at all, and yet had an important influence on the minds of those who were outside the building. We are thinking of Greek and Egyptian temples in which total darkness prevailed: the Gods were not in need of light! The people were not allowed to enter the building and it was sufficient to know of all the inscriptions and statues inside.

Still stronger proof of the sufficiency of unseen interiors and objects we find in the Egyptian graves, in which, shut away for eternity, the images of the dead, his possessions and the illustrations of his past and future life were stored.

In this chapter we have tried to prove that, with the aid of daylight as one of the most powerful tools, it is possible to bring the interior spaces in buildings to life. One of the most powerful tools indeed, but also one of the most difficult to handle!

We have mentioned only a few salient points in this art.

We know that many have been omitted, but it must also be remembered that there were whole periods and styles in which light played only an insignificant role.

Sometimes a beautiful application is the result of a logical construction or the proper use of a building material; at other times artifices and illogical constructions have been used in order to obtain fine light-space effects.

In the next chapter we will see how, in modern architecture, by reason of new building methods and often due to the less individual and characteristic use for which the buildings are destined, it is much more difficult to use daylight as part of the conception.

For artificial light we believe there are much greater possibilities.

44

Chapter III

THE PART PLAYED BY LIGHT IN THE ARCHITECTURE OF THE INDUSTRIAL ERA

In the previous chapters we have spoken of daylight giving shape and creating space in architecture.

In modern buildings the use of light is much more restricted to creating good working conditions; that means that attention is shifting from the building or the interior to the light required for our work. The interiors of the buildings of which we spoke in Chapter II were mostly intended to shut out the outer world. The buildings we shall now deal with are mainly places in which many people work. They always desire visual contact with the outside, so the windows must be so located and dimensioned that this communication with the outer world is possible.

The typical differences between past and present may be, briefly enumerated, as follows:

1. In the towns, lack of space very often forces the architect to accept situations that would certainly be rejected if there were more space and freedom.
2. Lack of space is also the reason for the tendency for buildings to become higher, and this again has great influence on the amount and the direction of the light penetrating through the windows and falling on adjacent buildings.
3. For high buildings, especially those with curtain walls, the elevations have to be as flat as possible in order to facilitate window cleaning. Protection against sunlight, for example by Venetian blinds and curtains, has to be erected on the inside, since sunshades on the exterior are expensive and unpractical.
4. Very often all the four elevations of a building have a uniform treatment regardless of their orientation.

   Bronze or aluminium window frames are necessary in order to reduce maintenance to a minimum (Fig. 44).
5. The windows are simple in form, the width is mostly chosen in accordance with the module of the rooms and partitions inside. This module is necessary because one must be free to divide or change the arrangement of the interior after the completion of the building.
6. The construction of modern buildings generally is such that the outer walls rarely support the weight of the building. Therefore the walls are light in weight and thin. This means that there are no deep jambs along the windows, the light enters with a very strong contrast between the glass and the frame.
7. It is obvious that for these buildings the sun causes most difficulties. All work in direct sunshine becomes impossible. In North America, where the climate shows much greater extremes than in Europe, it has become almost general practice to shut off the windows with venetian blinds down to about five feet from the floor, thus leaving a narrow strip of clear glass, allowing for the contact with the outer world.

   Here of course a high level of artificial light is essential, to be used all day long.
8. The necessary high levels of artificial light cause considerable heat which, added to the heat which in summer comes from the sun, makes it necessary to use large scale ventilation- and cooling-installations.

   These installations necessitate special constructions for the ceilings and also demand greater heights to accommodate ducts and ventilated light fittings.

All this also results in greatly increased building costs. Finally these costs decide to what extent the ideal lighting and climatic conditions can be realized.

45

Fig. 45. A good office, due to ample lighting, free outlook through windows, good furniture and plants and flowers.
Fig. 46. These offices, dull in colour, uniform in arrangement and furniture, and very uniform in lighting are an example of what should be avoided.
It is monotonous, inhuman and not conducive to pleasurable work.

As the result of these considerations it is clear that the use of daylight in modern buildings is no longer (see Chapter II) dictated by climatic circumstances. The construction has a strong influence on the size of the windows and the use of artificial light.

This means that the modern architect has fewer possibilities of applying daylight in order to obtain architectural and aesthetic effects, compared with his predecessors who could use traditional materials and constructions. This may, of course, be considered a loss, but we must not forget that a compensation may be found in the new opportunities in large installations for using artificial light, thus rendering those who work or live in the building independent of daylight. Such installations, which are almost impossible to use in existing traditional buildings, are easily applicable in new buildings where they can be planned simultaneously with the whole construction. They should not merely give utilitarian lighting but they should also make seeing easy and pleasant and create an agreeable atmosphere, give a degree of individuality to the space, and show the colours, materials and textures of walls, ceilings, floors and furniture in a harmonious way (Fig. 45).

If we think once more of the most impressive effects of daylight of which we spoke in Chapter II, we see that they were obtained by clever handling of light and shadow, of contrasts and directions of light. They show great differences in brightness near the windows and opposite them, and sometimes a ray of sunshine can have a most dramatic effect. It will be clear that, if we desire to obtain similar effects with artificial light, we will have to adopt completely new methods in order to obtain something of comparable value.

With one single lighting scheme for a great number of office rooms, or for other places in which a great many people have to work together, it will never be possible to obtain the variety and the vividness that we can obtain with daylight in buildings of traditional construction or in the home (Fig. 46).

We see already today a trend towards less uniformity in lighting, so that different parts of one room obtain differentiation in importance by accents in light and colour, according to their function.

Fig. 47. Working spaces with a high lighting level generally place too much emphasis on the ceiling. This may be compensated by using striking colours for walls and columns. (New York)

"surroundings of the task" are seen somewhat vaguely, without details and without colour.

Finally there is the periphery of the retina through which we cannot see any details or colours whatsoever. This periphery only reacts to strong contrasts of brightness and to movement which one can perceive out of the corners of the eyes. Its function is more one of warning. If we see a movement there we turn our eyes towards it in order to see what is happening. We find a similar state of affairs with animals. The protruding eyes of hares and of birds of prey are implanted on both sides of the head so that they can even see movements behind them.

Now if we look (for a very short moment) at a word on the page of a book while reading we see that one word clearly and precisely in line and colour. Around it we see the book (the task) with much less distinct details. We have to move our eyes in order to be able to read another word on the same page. We think we see more sharp details but that is only the result of very quick and often unconscious movements of the eyes, so that these details have been stored in the brain.

In the surroundings of the task the impression of colour gradually disappears and objects become very vague — one cannot recognize them any more.

The outer zone of the visual field, limited by the silhouette of eyebrows, cheeks and nose, gives us only information about great contrast and movement, through the periphery of the retina.

While studying the image in our field of view we often have to use the terms: the centre of attention, the eye task, the surroundings of the eye task, the background and the outer zone of it.

In order to define these notions more precisely we made a few experiments.

The centre of attention is very small and corresponds with the fovea of the retina ($\pm$ 2°).

The eye task covers that part of the visual field we can see clearly and sharply with our eyes moving and our head kept still. That implies that we perceive the eye task through the most sensitive part of the retina. If we look at an opened book of normal size at a distance of about 40 cm (normal reading distance) we get an approximate

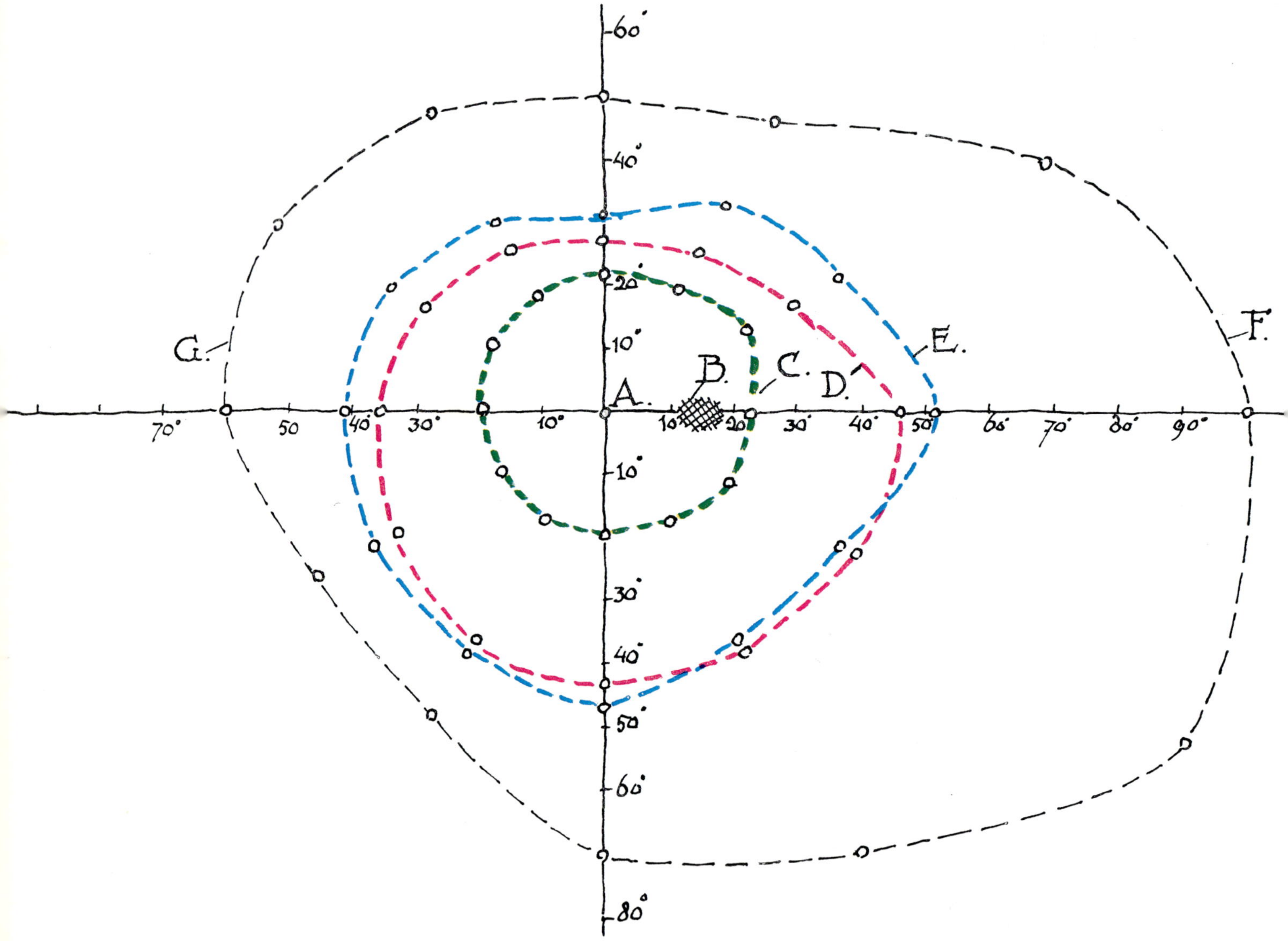

Fig. 49. Diagram of the retina of the human eye. The sensitivity to light and colour is not uniform over the total surface of the retina. The centre, the fovea, is the most sensitive. Around this centre is a zone in which cones and rods are mixed, but outside a cone of roughly 2 x 45° there are only rods so that there no colour sensation exists, and only great differences in brightness and movements are perceived.

idea of the solid angle in which the eye task is included. We have found for this solid angle an elliptic cone of 8° above and below and 10° to left and right of the line of sight.

The surroundings can be described as that part of the visual field in which we can perceive some details and colours without moving the eyes. We found the dimensions of these surroundings as follows: If we want to see an object as a whole (e.g. a painting, a piece of furniture, a person) we automatically choose a distance for our observation which depends on the largest dimension (height or width) of that object. If, for instance, the horizontal dimension is the biggest (e.g. a painting of 1.50 m wide and 1 m high) we choose a distance of about 2.5 times that distance for observation ($2.5 \times 1.50 = 3.75$ m). For a person 1.80 m high we choose a distance of about 3.2 times the height.

This means that we can limit the surroundings to an elliptic cone of about 18° above and below and 22° to the left and the right of the line of sight.

The total field of view, including the outer zone, is very large, about 100° to left and right. It is limited by our eyebrows, cheeks and nose so that we might indicate its approximate shape as in Fig. 50. We will call all we can perceive between the outer limit and the limit of the surroundings the background. We have dealt with the character and the sensitivity of the different zones in the retina and the zones we distinguish in our visual field, because we want to make clear that there is a close relation between those two.

In Fig. 50 we have once more drawn the main part of the visual field, the eye task and the surroundings and part of the background. Here we find that the limit of the sensitivity to colours in the retina is to be found just outside the environment!

The dotted lines give (Sewig. Handbuch der Lichttechnik, Vol. I page 257) the limits outside which we do not see red (the outer circle) or green (the inner circle).

A simple test can confirm these observations.

If we hold up a few fingers or a coloured object to a person looking at a point some 40° away from that object he cannot see the number of fingers or the colour the object.

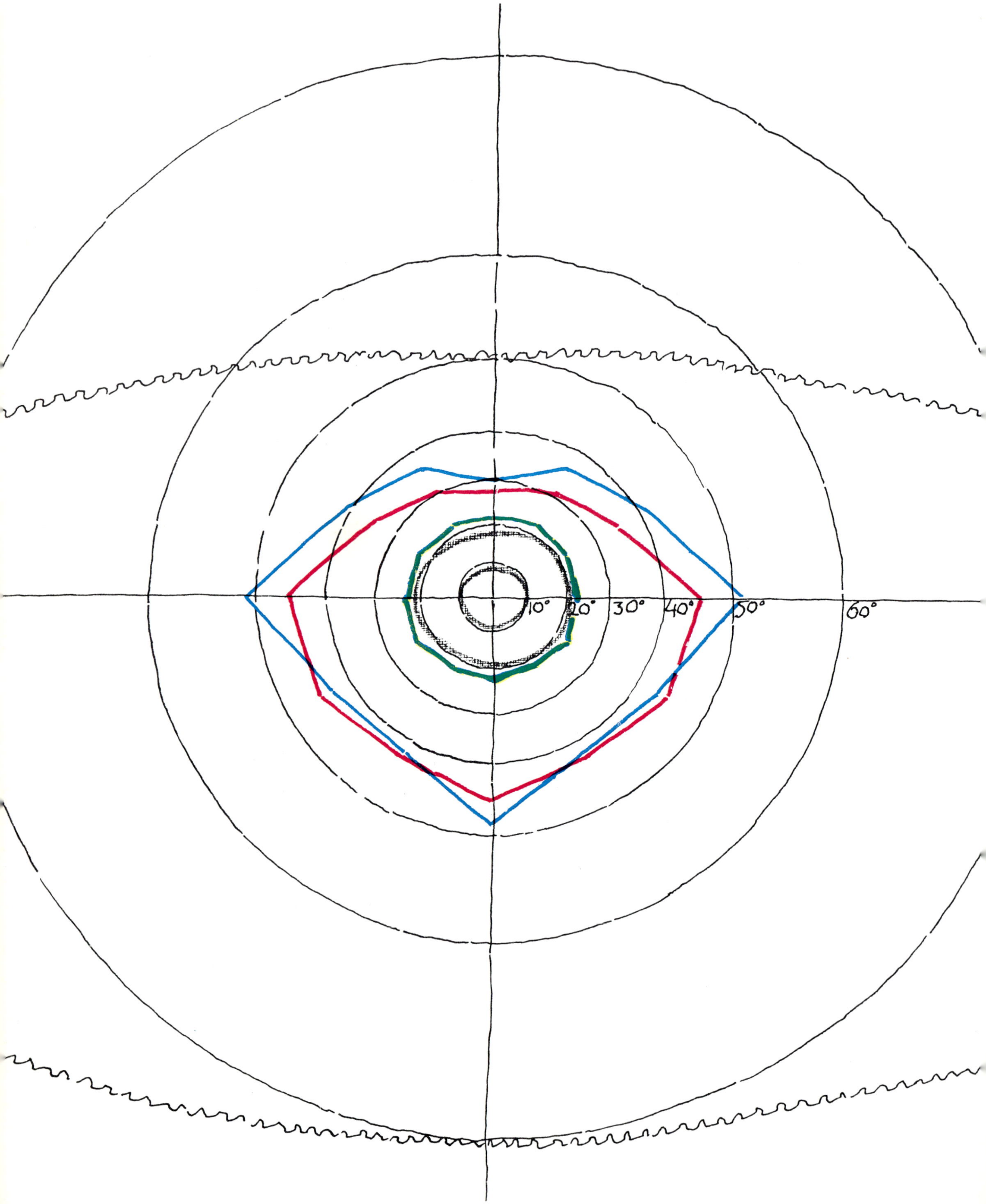
10°
20°
30°
40°
50°
60°

Outside these limits we are also less sensitive to great differences in brightness. For that reason the brightness of the sky in a landscape, or of the windows in a schoolroom do not disturb us. Moreover, direct glare coming from these directions is much less disturbing than that originating nearer to the line of sight. If we want to see an object or a scene in detail, we have to scan it all, with swift movements, so that we have traversed all the important details with the fovea and the retina immediately around it.

All these elements of information are assembled in our brains to form one image, which is then more or less confirmed by the vague outlines we see when we again concentrate on the centre of the scene. If we notice a great contrast in brightness or a movement in the outer limits of the visual field with the periphery of the retina, we have to turn our head in order to see what is the cause of the distraction. These eye movements, which are very quick and complicated cause different light sensitive elements to function with each movement. This is essential since rods and cones lose their sensitivity after a split second, so they constantly need changes in exposure in order to renew their sensitivity by a biochemical reaction.

One can make a simple but interesting experiment to illustrate this.

If we fix our line of sight as completely as possible by concentrating on a tiny little spot without moving the head or the eyes, the environment and the background become one indistinct blurr after a few seconds.

All the information we receive through our eyes does not interest us in the same measure. According to this interest each piece of information will be digested and either retained or rejected and forgotten (see also Chapter IX).

They may constitute an affirmation or a support for our thoughts, or a distraction; they may also have no relation whatever with the things that occupy us at the moment, and thus provide a neutral background for our work or our thoughts.

In the following chapter we shall see how we can arrange the visual field in such a way that the impulses coming from it will be favourable to our concentration and our work.

Fig. 50. The central part of the visual field in which the eye task is restricted to a very small focal area (one word on the page of a book); the surrounds of the task in which we see details and colours (though slightly out of focus) and the background.

Chapter V

THE DESIGN OF THE VISUAL FIELD

It was seen in Chapter IV that usually the eye task, that is to say the object we observe carefully during our work and which we want to see in detail, occupies only a small part of our visual field. The details we see sharply are still smaller, they may be one or two words on a printed page looked at from a normal reading distance (40 cm); or that part of a piece of knitting in which we want to count stitches; or the upper part of the body of a violonist standing at a distance of 20 metres on a platform; or that part of the road on which we see a car approaching at 300 metres distance. This part of the eye task is not wider than a solid angle of $\pm 2°$. If, as stated in Chapter IV, the whole eye task which we can see with fixed head and moving eyes is included in a cone of 8° upwards and downwards and 10° to right and left, the part we can see with the greatest precision is only 1/500 to 1/1000th of it. Now it is obvious that since the whole visual field contributes to the total impression we obtain when looking at our work, the whole composition of that visual field has an influence on the way we see things.

If it is agreed that during work easy and quick seeing is important, the total image in the visual field becomes important for two reasons.

In the first place the image can help in seeing the eye task.

For instance a ballet-dancer in white becomes clearly visible against a dark background, but a reading task becomes more difficult if the printed paper lies on a shiny black desk in which the light fittings and lamps are reflected.

So the environment and the background can distract us or help us while we are looking at a task. Distraction may be caused by glare, movements or excessive contrasts.

In the second place seeing becomes more easy if the environment can add to the concentration of the observer.

Quick and easy seeing demands two main conditions:

— It must be possible to see the characteristic details of the eye task clearly, the shape, the colour, the size, the shadows and the texture. To ensure this it is sometimes necessary to apply special shadow effects, high lights, high lighting levels or a special colour of the light.

- The second condition is to have, during work, a visual field that can help us concentrate on the task.

The present chapter deals with this condition. How can we facilitate easy and concentrated seeing by a special arrangement of the visual field?

The image in the visual field is always composed of three elements: line, colour and brightness.

Every image we see, a painting, a landscape or an interior, consists of these three elements.

We can study them separately and try to discover the role each plays in the impressions we receive from what we see.

Let us start with the lines.

Lines always have direction and our eyes have a tendency to follow their direction. In our field of view, of which we see only a small part distinctly, we are always concious of the main lines in the total field. Those lines may converge towards the task, they may more or less run concentrically around it, or they may have no perceivable relation to the task (Figs. 51A and 52).

If we want to indicate a word on a page we can do so by either drawing a circle around it or by an arrow pointing towards it. The influence of lines is strong, as for example that of the white lines or the row of white poles guiding the traffic along the road.

The rows of houses, the kerbs, verges all pointing in perspective towards the visible end of the road help us to concentrate on the eye task. For night vision the street lights have to be carefully positioned, so that the line of the lights indicate the direction in which to drive (Fig. 51B).

Parallel lines emphasise a direction still more strongly. We may think, for instance, of the columns in a Gothic church in which the multiplicity of vertical lines guide our eyes upwards (Fig. 53). But the concentric lines of the vaulted roof of a refectory keep the eyes of the monks concentrated on their prayer books (Fig. 54).

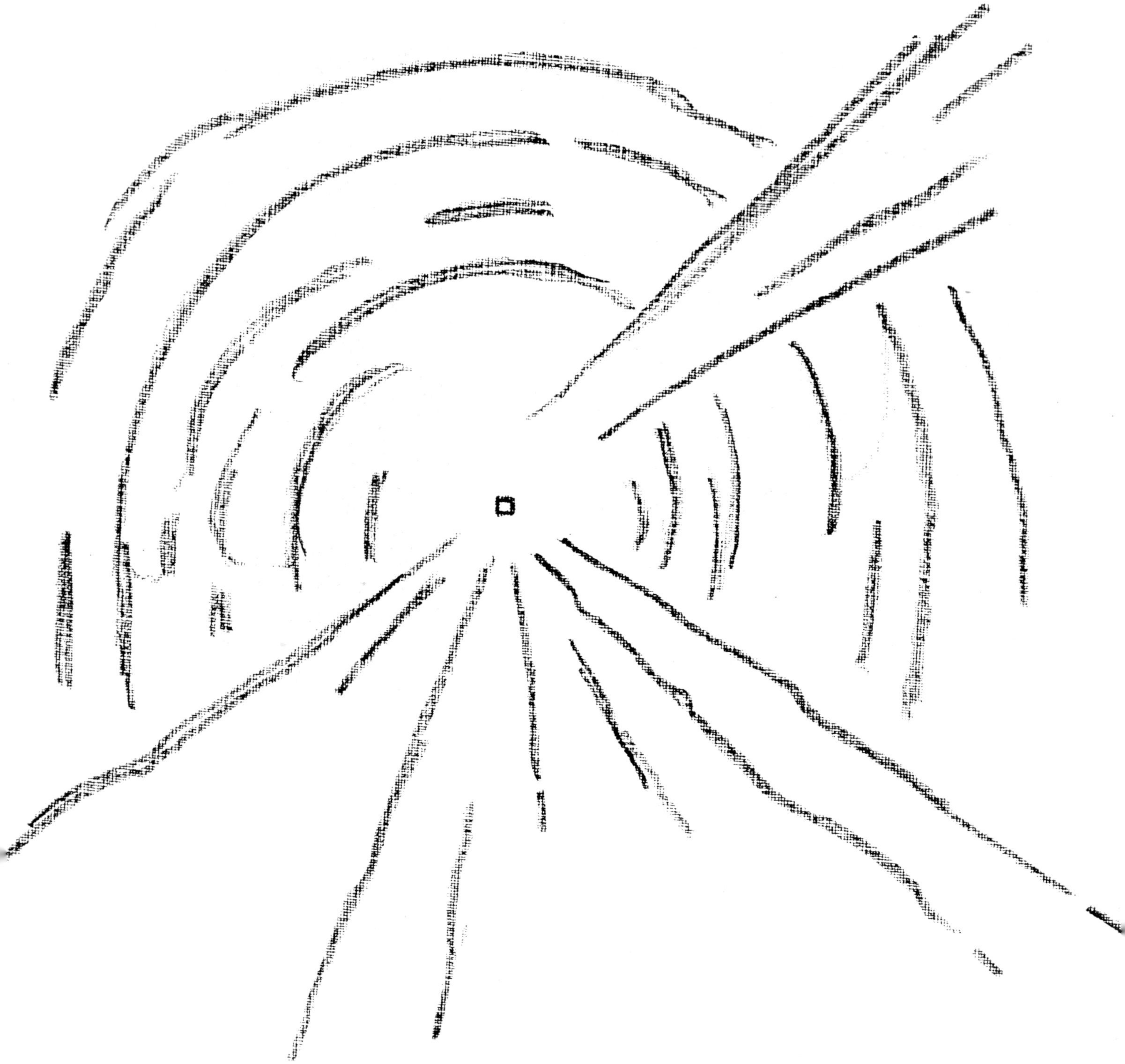

51A

Fig. 51A. Concentric lines and radial lines arranged around the eye task draw and hold the attention focused on the work.
Fig. 51B. If we drive a car along a road we find that the white lines along the verge and the converging lines of the light fittings on their standards facilitate the concentration of the driver on his task.
Fig. 52. Erratic lines around the eye task may be confusing and distracting in any case they do not help us to perceive with ease and concentration.

51B

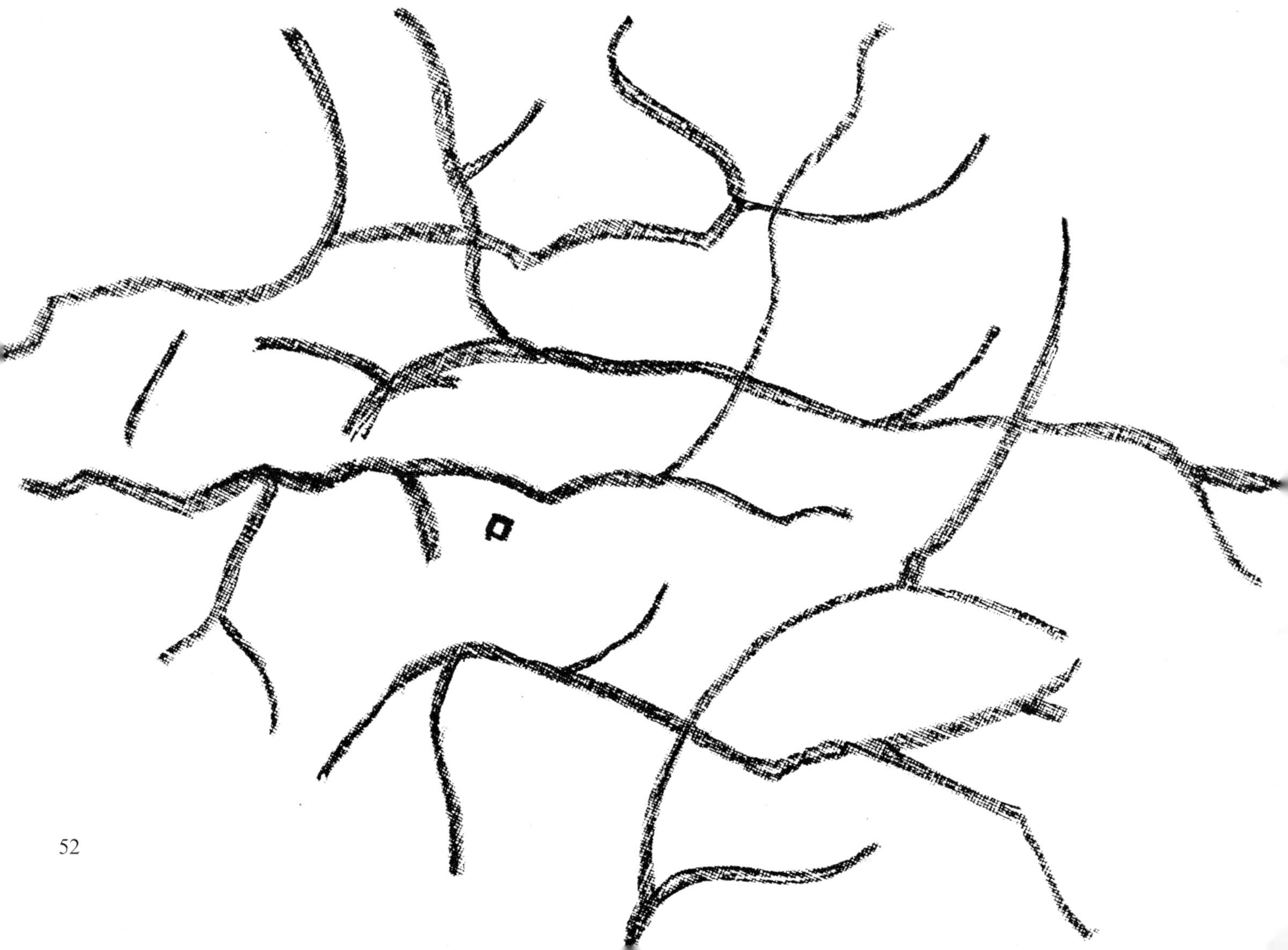

Fig. 53. All vertical lines of the columns in this cathedral guide our eye upward. ('s-Hertogenbosch)
Fig. 54. We look up in a cathedral, following the direction of the vertical lines of the columns; but in this refectory the concentric lines of the arches strongly emphasise the horizontal direction. Here the monks would walk up and down and read a prayer-book. (Fontenay)

Fig. 55. The erratic lines mentioned in Fig. 51B occur as dark shadows of branches and trees on a sunny road. It is common experience that it is tiring to drive along such a road.

Lines or combinations of lines that have no relation to the direction of our vision may have a confusing and distracting influence. As an example we may take a sunny road on which an erratic play of shadow lines from trees makes concentrated driving very tiring (Fig. 55).
We give here two simplified diagrams, the first showing how lines around the task

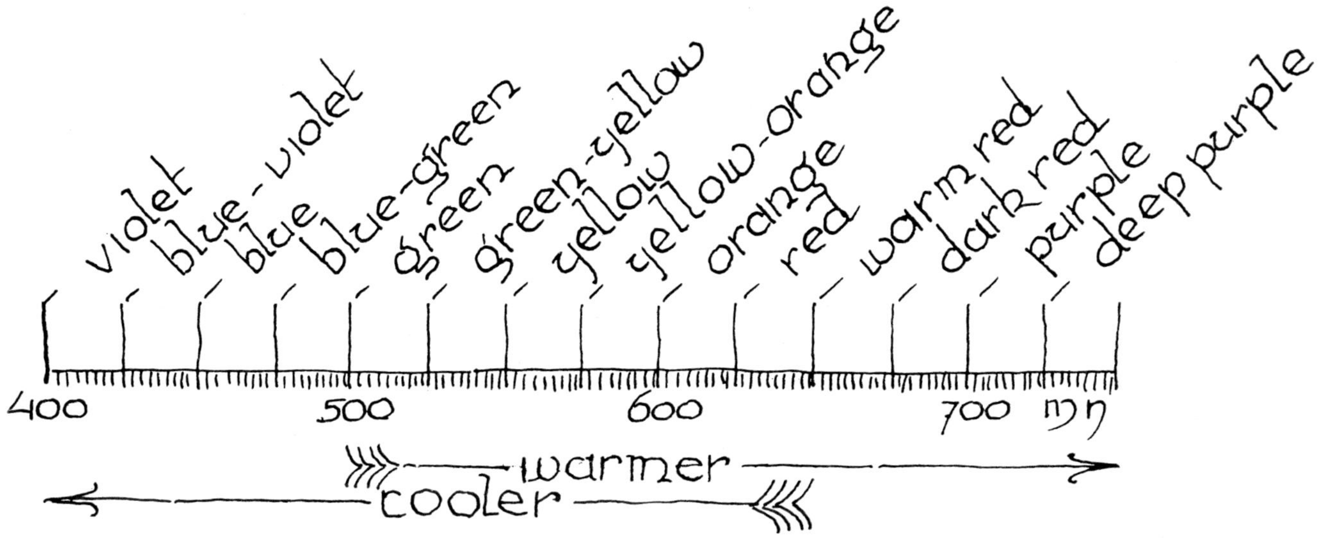

Fig. 56. The visible light rays have wavelengths between 400 and 760 nanometres, that is between red and violet. We call every colour that has a longer wavelength than another the cooler colour, and every colour of a shorter wavelength than another the warmer colour. So green is cooler than red, blue cooler than green, but orange is warmer than yellow, green warmer than blue.

(the black spot) may help to concentrate our attention on it while the other, in which the lines have no relation to the task, make concentration more difficult. (Figs. 51A and 51B).
For our work and other eye tasks of long duration we must take into consideration this positive or negative influence of the lines in the field of view and, by the judicious use of lines in the visual field, facilitate the ease and concentration of our perception.
We will now consider in a similar way the action of the colours in the visual field. The lines have significance by their shape and direction; the colours are differentiated by their wavelengths — the shorter wavelengths correspond with what we call the cooler colours, the longer wavelengths with the warmer ones.
We will start from the thesis that the warmer colours have more attraction than the cooler ones; and that, as the name indicates, they have a more lively and stimulating effect on us.
All wavelengths in the visible spectrum lie between 400 and 760 nanometres. We might draw a line at 550 nanometres (about yellow-green) and call all the shorter wavelengths the cool colours, and the longer wavelengths the warm ones. But it is better to say that any colour of a long wavelength makes a warmer impression than colours of shorter wavelengths. This means that we regard yellow-green as warmer than blue-green; that green is warmer than blue; and that yellow is cooler than red or orange (Fig. 56).

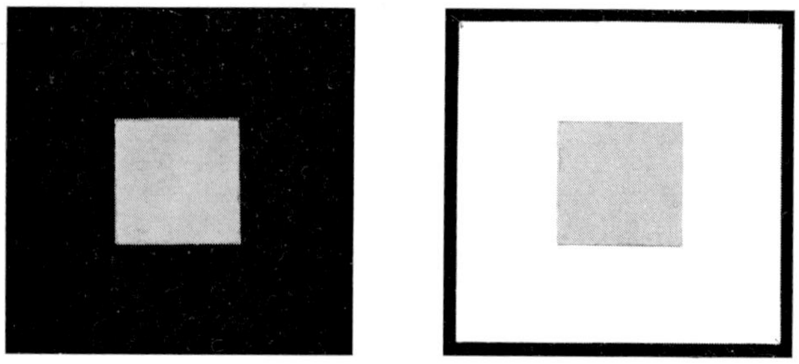

Fig. 57. Luminance is a physical conception, it deals with the amount of light coming in a given direction from a certain surface. Brightness is the subjective impression we get of that luminous surface. The luminance of these two small squares is identical but the brightness of the square with the black background is greater than the brightness of the square with the white background.

Fig. 58A. If the task is a bright centre, surrounded by surfaces that gradually become less bright the further they are away from the centre, the attention is drawn by that task.
Fig. 58B. In order to prove the effectiveness of correct brightness distribution around the tasks the brightnesses have so been changed that the task is in the darkest position. It is obvious that Fig. 58A is much more favourable for concentrated perception.

If we now think of the eye task as surrounded by elliptic zones like we did in Chapter IV, and if we could give colours to these zones, it would seem logical and effective to give the inner zones a warmer colour and choose gradually cooler colours for the zones more remote from the task. When doing this, however, we may not forget that the effect of a colour depends on its saturation and on its purity.

A red paint mixed with white (pink) is much less active than the pure red. Similarly a red paint mixed with black, to give a darker red, is less active than the pure red. Saturated colours are rare in general, and since they have a strong impact, should be used sparingly, that is to say occupying small surfaces. An example is the telling effect of a few red roses in a vase.

If we apply strong colours it is necessary to consider carefully the background against which they will be seen; a strong contrast may be annoying.

Let us consider a few examples:

When reading, it is preferable for the white paper to lie on a warm brown wooden table, which in its turn stands on a dull green or blue floor against a medium grey wall.

A table with an intense green tablecloth standing on a red carpet would certainly distract our attention.

In a concert hall we may accept a singer in a red dress, standing against a grey or green background, but it would be strange and wrong if she wore a grey or a green dress standing in front of a red curtain.

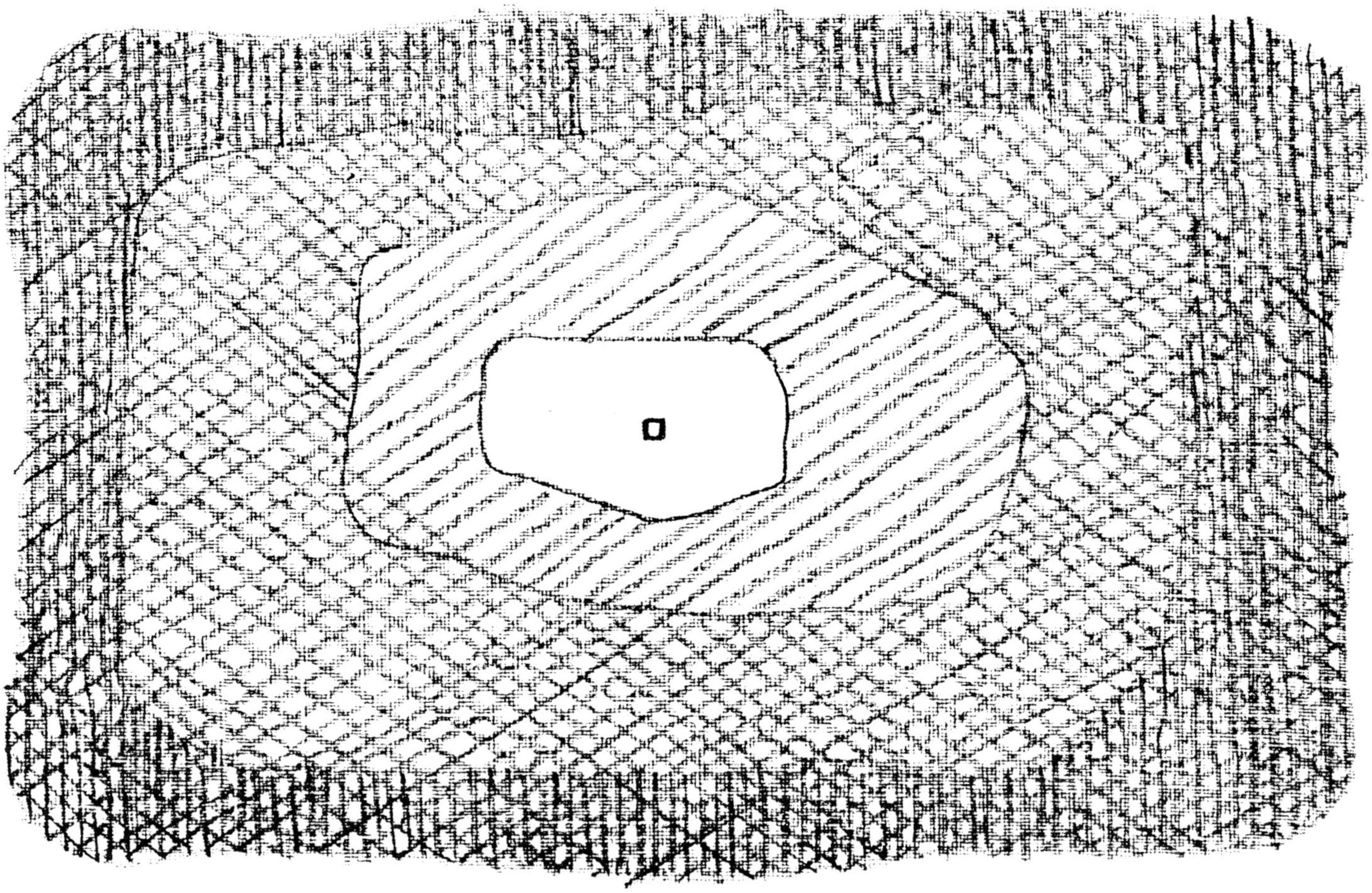

58A

58B

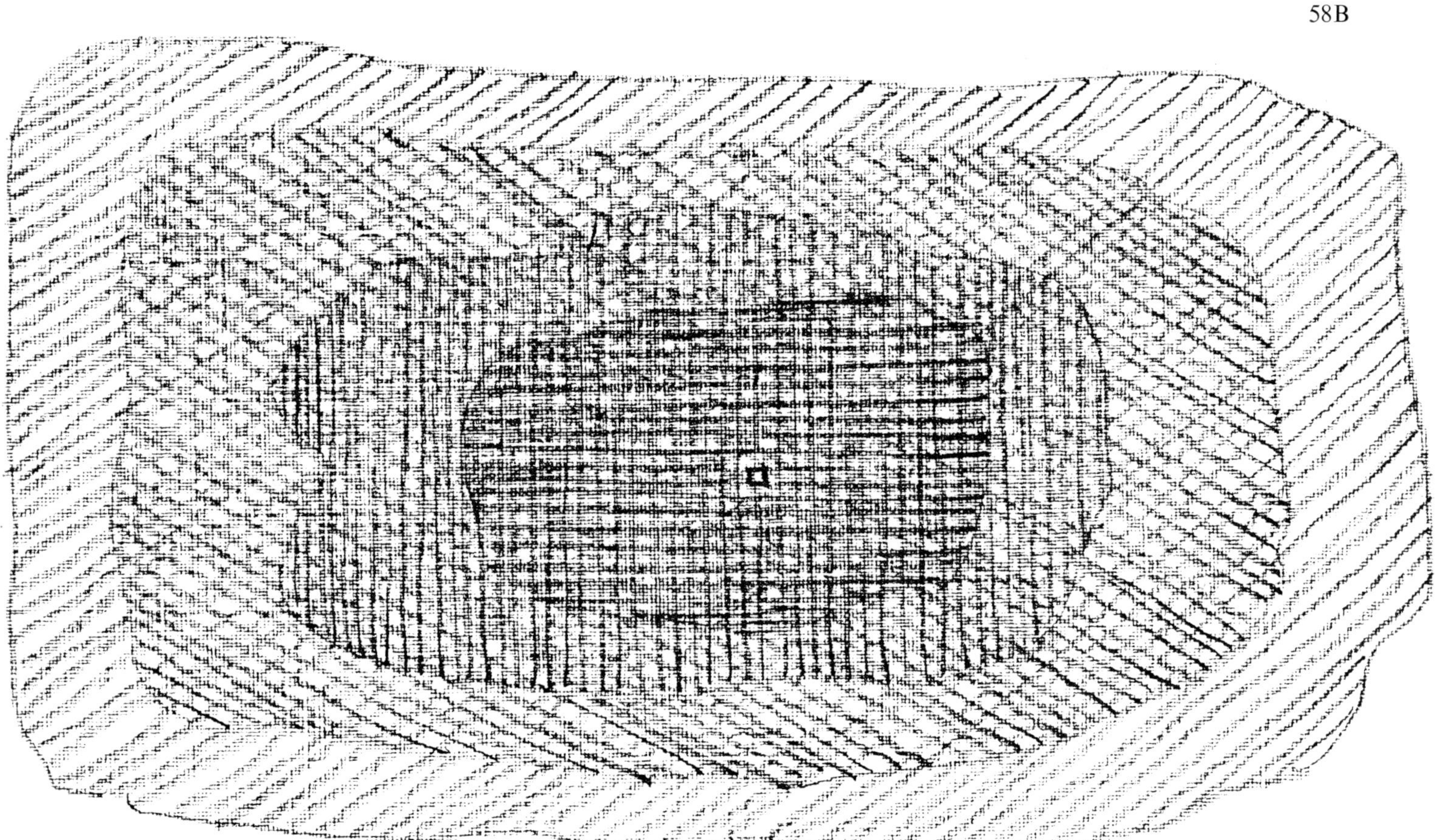

We can also give a culinary example. Red lobsters and red tomatoes on a background of green salad leaves make a very striking and attractive dish. If lobsters, after being cooked, stayed a blackish green they would be much less desirable!

So far we have dealt with the positive influences of colours in the field of view on the ease of seeing and on our concentration. This influence may, in many cases, seem undesirable.

All colours, if mixed with white or black, soon become almost neutral and can hardly be called warm or cool. Such tints, together with white or grey are suitable for the many cases in which we do not wish any influence by colour on our concentration.

We now have to deal with the influence of the brightnesses in the visual field.

Brightness may be considered as the personal impression the observer obtains from the lighter or darker surfaces he sees. Brightness is not something we can measure, because it is relative and subjective. Objective brightness is called luminance and is, of course, measurable. Brightness is influenced by the amount of light falling on a surface, by the reflectance of that surface and by the brightness of the surfaces around it.

If we place a piece of grey paper on a black table it becomes brighter than if it lies on a white table, although the luminance will stay the same (Fig. 57).

Now it is common experience that the brightest surfaces in our visual field attract our attention most.

This means that our concentration is increased if we accentuate the essential part of our visual field by giving it the greatest brightness. The visual task, being the essential centre of our attention, should be the brightest, and the immediate surroundings, the background and the periphery should get progressively less bright. This again is made clear by the accompanying diagram in which the task in the centre is the brightest and the successive zones around it become gradually less bright (Fig. 58A).

If we revert the brightnesses, and the task lies in a dark zone surrounded by lighter zones, it is difficult to concentrate on that task (Fig. 58B).

A few practical examples illustrate the validity of this thesis.

- The dancers in a ballet will always wear white or very light coloured costumes and dance against dark backgrounds. The prima ballerina will be given extra emphasis by spotlights concentrated on her.
- A tunnel in a landscape is an unobtrusive little black spot amidst a vast composition of light and colour; but if we are in that tunnel, the opening through which we see the brightness of the sky, surrounded by dark walls, is the only point to which we direct our atention.

- Spotlights used in shop-windows are useful for accentuating certain goods. For those items the window dresser will, of course, choose light-coloured objects. A black object would not reflect sufficient light.

If at a certain moment an object becomes important, we make it interesting by giving it greater brightness.

At dinner-time the table becomes important, and so we spread a white cloth on it.

Napoleon, when about to lead a battle, was dressed in white, and rode a white horse.

At a wedding, the bride is the centre and so she wears a white wedding dress.

We have successively spoken about lines, colours and brightnesses. It is obvious these never occur separately and that the three combined in the best way can have a much stronger effect than if used separately. We can also imagine combinations in which the three factors have opposing effects, which of course reduces the viewer's concentration.

In order to show how the three factors can co-operate, we reproduce diagrams covering the central part of the visual field of the observer. Each diagram consists of lines, colours and brightnesses.

- In the first diagram the lines run in three concentric ellipses around the task and two diagonals point at it. The colours, starting in the centre, are brownish red, grey, green and dark-grey. They make the centre warm and attractive and the surrounds cool and passive. The task itself has been indicated as a white square with a small black square (black characters on white paper) and the brightnesses are highest in the centre (Fig. 59).
- The second diagram is, but for the colours, the same as the first. The colours have been reversed, the cool colours in the centre with the warm ones surrounding them (Fig. 60).

If we compare these two diagrams, the first seems to be most adequate for concentration; the second distracts the attention from the central part of the visual field due to the warm red zone outside the centre.

The first could be a good example for a visual field while looking at a task with the head kept still and eyes moving (e.g. for reading a book). It would make seeing easy and concentrated.

In pratice we do not constantly keep our eyes fixed on the task. From time to time we look up from our work and by looking at a more distant object relax our eye muscles.

This means that we often have two centres of attention. In the third diagram such a situation is shown (Fig. 61).

In the lower centre the task is the reading or writing (black on white) whilst the upper centre on which we concentrate while relaxing shows red on white (e.g. a bunch of red flowers in a white vase).

For other visual tasks different diagrams can be prepared.

If a number of observers sit around a table each observer has his own field of view. The diagram must be valid for all the observers, which means that it must have horizontal parallel zones of brightness and colour. If we think of a meeting round a table, or guests sitting in a restuarant we find two main zones of interest; the zone in which we see the faces of the persons across the table and the zone of the papers or of the meal on the table.

These two zones should again be the warmest in colour and the brightest; the zones of the floor and the ceiling should be cooler and darker.

Fig. 62 is such a diagram for a restaurant, with white linen on the tables, well lit by lamps in warm-coloured shades, spreading a warm light on the faces of the guests. Walls and ceiling remain in semi-darkness and have darker and cooler colours.

If, however, we imagine a situation with a white ceiling and indirect light, a light grey carpet as shown in Fig. 63, the ceiling would be distracting, creating an atmosphere quite unsuitable for a restaurant.

In a church the situation is again completely different. Here we want the main emphasis in the centre of the nave with the choir and the altar, for which we need the greatest brightnesses and the warmest and most vivid colours. We do not find an eye task claiming our full attention, we let our eyes wander upwards, following the vertical lines of the columns. A scheme as given in Fig. 64 seems to provide the right answer. If we reverse the brightnesses and colours as shown in Fig. 65 the essential parts of the church will lose much of their visual attraction.

We have been discussing the composition in a visual field which we will see during lengthly periods of work or relaxation. All elements which contribute to this composition should co-operate in order to obtain the most favourable conditions for seeing the eye task.

Not only the architect, the decorator and the lighting engineer are responsible, but also the maker of the paper on which we write or from which we read; the designer of the furniture who will decide about the colour and the texture of the desks and table tops; and finally the man in the purchasing department of the firm.

Let us quote here a seemingly unimportant example.

In the firm where we worked for a long time a half-yearly telephone directory is distributed, each time in a differently coloured cover. At one time that cover was a vivid red. This colour next to the telephone on our desk proved to be annoying and distracting.

We have now built up a theory, partly on logical arguments partly on the reactions we have observed when looking at our environment. It is difficult to define these reactions to analyse and to measure them seems impossible.

For this reason we have tried to find confirmation of this theory in the work of painters — people who lead a specially active visual life.

We will therefore analyse the compositions in line, colour and brightness of a number of famous paintings, dating from various periods.

"Gallant Scene" by Jean Honoré Fragonard (1732-1806). This painting from the Wallace collection, London, strikingly demonstrates the co-operation of the three elements — line, colour and brightness — as a result of which our eyes do not hesitate for a moment when looking at this charming picture (Fig. 66).

The clear and warm colours, pink, red and gold of the girl on the swing, surrounded by the gradually darkening and cooler colours of grey, green and blue show how our eyes are attracted more strongly by warm than by dark and cool colours.

The lines in the picture — the cords of the swing and of the man pulling at the swing, the outstretched arm of the impertinent onlooker — all point towards the visual centre. The tree trunks, branches and foliage together make circles accentuating this centre.

"The month of July" a typical Dutch landscape by P. C. J. Gabriël (1823-1903) in the Rijksmuseum in Amsterdam.

This painting illustrates in an almost ideal way how a great many painters compose their paintings along principles analogous to those put forward in our theory.

Fig. 64. In a church there is hardly an eye task, but we often feel an urge to look upward following the vertical lines in the building (see Fig. 53). There is, however, no question about what the essential line of sight is. It will be directed towards the centre of the church, the altar and the priest.
This part of the church should be made the brightest by light and colour, and it should be warmest.
This diagram is an example of how the line, brightness and colour composition should be and generally is arranged.

Fig. 65. In this diagram the positions of the colours and brightnesses have been changed. The centre of the church is no longer attractive and the brightness on the two sides distract attention from where it should be concentrated.

Chapter VI

THE RHYTHM OF NIGHT AND DAY

All living creatures are strongly influenced by the rhythm of the sun and the moon, of the seasons, of day and night, of ebb and flow. Mankind is equally influenced by these rhythms.

Plants show remarkable reactions to light. Many plants obtain a stimulus from the lengthening of the days and from an increase in light intensity. Spring flowers such as daisies, tulips and cherries are good examples.

The nursery-man makes good use of these phenomena; for instance, forcing tulips and other bulbs to flower, by lengthening the day for them with electric light, which often entirely replaces the scanty wintry daylight.

There are other plants such as chrysanthemums that bloom in autumn and start opening their buds when the days are shortening and they receive less light. Here again the nursery-man makes use of this property. In order to retard the flowering till after Christmas, the day is lengthened artificially by supplementary electric light. When this extra is stopped the flowers start to open.

With animals we find similar phenomena. Fifty years ago eggs became scarce and expensive in autumn, the hens started to moult and did not commence to lay eggs again until spring, so that by Easter eggs were abundant and cheap. Modern practice is to lengthen the days for hens artificially by employing artificial light after dark. This keeps the fowls active so that they keep laying eggs all through the winter. In all chicken farms the lights are switched on in the poultry houses at sunset in the winter.

The sun and the long summer days also have an activating influence on man. The cold dark days of winter often depress us. The cycle of day and night also strongly influences us. In the morning we wake up after a night of rest and sleep. The sun is up, the daylight is abundant and of "cool" colour. Man starts his day full of energy and ready for activity. For many the daily walk to work in the fresh morning air is a powerful stimulant.

At the start of our days work we like to have plenty of light and cool and light colours around us.

When the sun rises higher the light becomes more profuse and warmer in colour.

During the second half of the day much of our energy has already been spent; we may become somewhat weary and tired; the work advances less quickly and we begin to long for rest and relaxation. This longing is accompanied by the gradual decline of the sun and the consequent reduction of the light which, with the setting of the sun assumes a warmer colour.

All our life we experience, day by day, these regular changes in quantity and colour of light, accompanied by the daily fluctuations in our energy and activity, and subconsciously we combine the two. Now the sun is setting, the evening sky is becoming red and our concentration and energy are slowly ebbing away.

If we have to continue work we must supplement the dwindling daylight with a high level of cool artificial light, so creating a new stimulant for our work, by bringing back the active atmosphere of the morning. At the end of the working period we feel tired and we long for rest and relaxation, and in our homes we therefore desire less light and warmer colours. These we find not only in the home but also in restaurants, dance halls, bars and the theatre.

Our intellectual activity will usually diminish and our emotional life assumes a bigger role. We engage in reading, conversation, music, games, watching films and plays in the theatre, and enjoy all things which call for an intimate and warm atmosphere and lighting.

Roughly one might illustrate this cycle in a diagram; the daylight starts with sunrise, has its maximum about noon and diminishes again till sunset. In that period of an average of 12 hours the colour of the light gradually changes from cool to warm. This we indicate in dominant wavelengths in the diagram of Fig. 71.

In the afternoon when the amount of light, and with it the lighting level, falls noticeably, we will need supplementary artificial light in sufficient quantities and of the right colour in order to be able to continue doing our work with the same concentration.

After work, the total requirement of light becomes much smaller. When it is dark outside we enjoy rest and relaxation, quite contented with warm light (mostly from incandescent lamps) in quantities that are only a fraction of the light we need in daytime as a supplement to the failing daylight.

It goes without saying that mere variations in colour and quantity of the light are not sufficient; we must decorate the rooms in which we work or relax accordingly, creating, usually by the use of colour, the requisite atmosphere and mood.

In the rooms we occupy after working hours we employ red, brown, orange, yellow freely, but in the daytime workspaces white, grey, green and blue tones predominate. The light employed during the day to supplement the daylight is usually provided by fluorescent lamps, not only because these lamps have a high light output but also because they can produce light of much cooler colours than incandescent lamps.

We have now described the influence of the "rhythms of light" in our daily life for latitudes between 40° and 60°. However, in these northern countries conditions are completely different from those in subtropical and tropical climates, where, as a result, the activities by day and night are also very different.

In the tropics where seasonal changes corresponding to summer and winter in temperate zones are unknown and where day and night have the same length the whole year through, the temperature between noon and three p.m. generally calls for a siesta, and the appreciation of light is quite different.

Intuitively the amount of light is adjusted in relation to the temperature, so that the maximum illumination values are generally lower than in the colder countries. The colour of the light should be as cool as possible so that a bigger market for the cool fluorescent blueish light exists in the hot countries than in the milder climates.

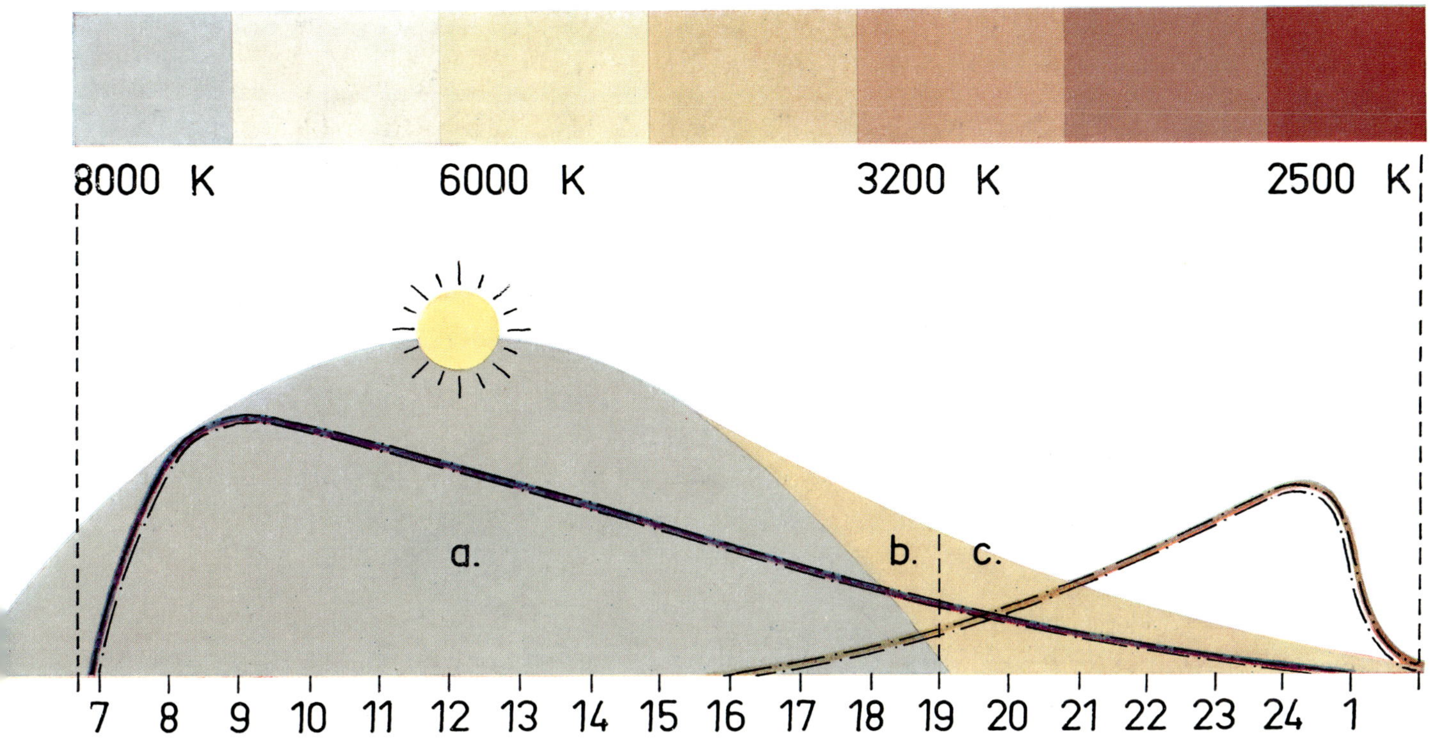

Fig. 71. The colour temperature of daylight lies between 8 000 and 4 000 K (a). Supplementary artificial light is between 5 000 and 3 000 K (b). Evening light (c) between 3 000 and 2 500 K. The blue line indicates our daytime activities, the red line stands for our emotional occupation.

Chapter VII

HOW MUCH LIGHT DO WE WANT FOR OUR WORK?

The human eye can see under a wide range of lighting levels. This adaptability is easily recognised by considering our powers of perception under conditions changing from full sunshine to moonlight.

When looking at an eye task during a longish period we find, however, that an adequate lighting level makes seeing easier, or more precisely, that normal reading, for instance, calls for less effort or is less tiring.

In general, seeing an eye task becomes easier the greater the amount of light used. For many years now lighting experts have tried to find the optimum conditions for seeing an eye task, and the lighting level necessary for that optimum.

The idea was that the proper lighting level would yield the highest efficiency for the eyes and that exceeding this lighting level would represent a waste of light and money. It is obvious that eye tasks which differ in contrast and in detail call for different lighting levels.

It has also been proved that older persons need more light than younger ones (5 to 10 times more!) in order to have the same visual acuity, so that an absolute level for a given eye task cannot be specified (Fig. 72). People tried to do so, however, because the lighting industries and the power companies were interested in selling more light and the consumers were keen on using a minimum of current! These are, however, as we have already indicated in previous chapters, a number of other factors which influence our seeing and render it easier or more tiring.

These factors induce negative factors such as glare, specular reflection, small contrasts between the object and the background, and positive factors such as clear details and well designed shapes (typography), easy concentration (see Chapter V), co-operative patterns of brightness and colour in the visual field and a pleasing environment.

It is interesting to observe how these positive or negative factors which influence our visual perception increase or decrease at higher or lower lighting levels.

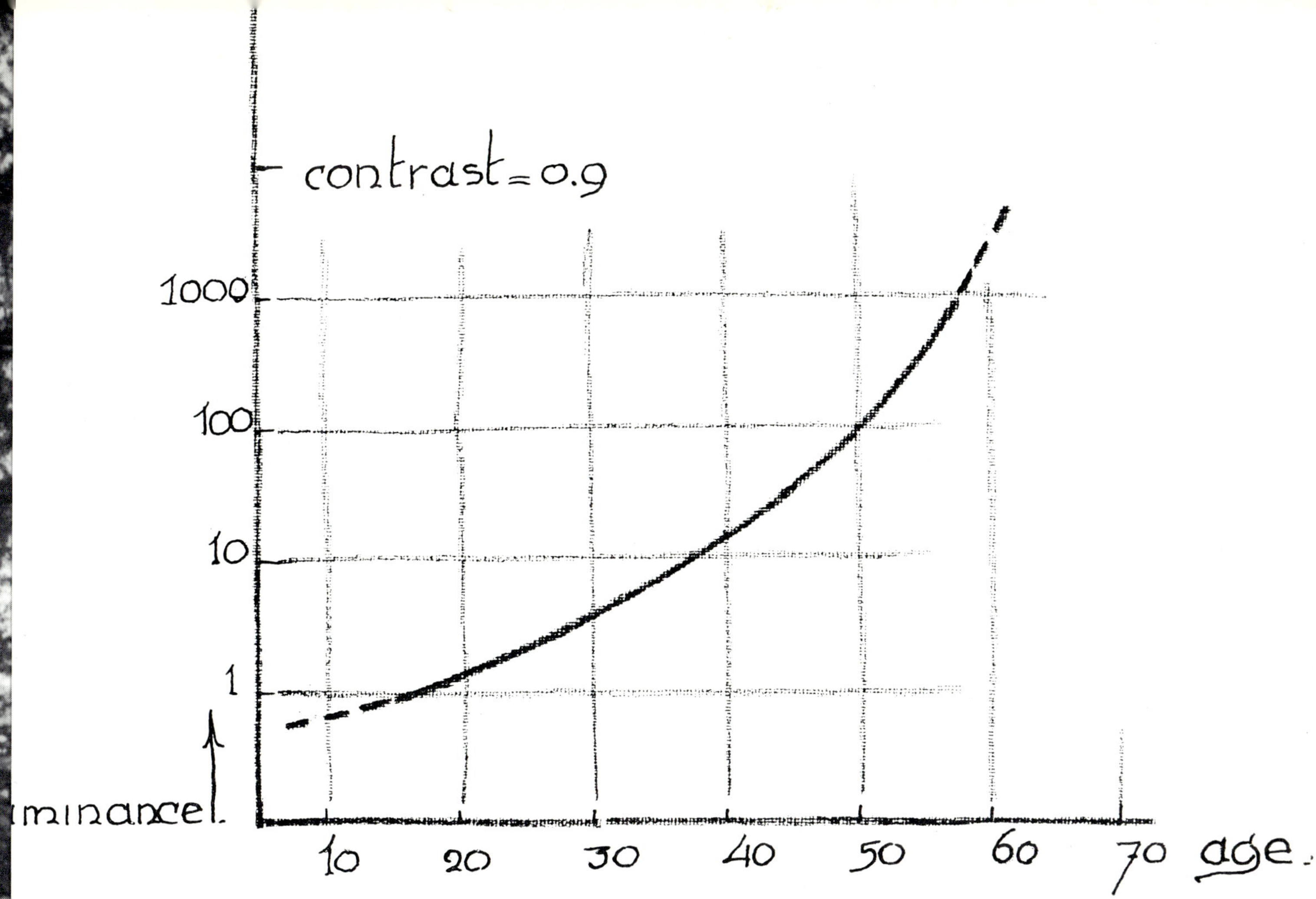

Fig. 72. This curve shows that persons of different ages need different luminances in order to easily see a given object against a given background. For the task a man of 20 years can see easily (at 12 cd/m^2) a man of 40 needs ten times the light and a man of 50 about 100 times the light! (Dr. G. J. Fortuin, 3rd Congress of the European Society of Ophthalmology, Amsterdam 1968)

Glare can be very uncomfortable when, looking constantly at an eyetask, one sees simultaneously bright lighting fixtures or light sources. This happens for instance if one drives along a well lighted road and passes other vehicles with glaring head lights; or doing office work in a vast room in which the whole ceiling is studded with many light sources of great luminance.

But this very glare may be very agreeable in the festive illumination of a dance hall, or when looking at a firework display, or during a walk in a green valley with sunlit snow peaks around it (see figs. 73, 74 and 75).

Between 1920 and 1930 the use of artificial light was still very limited. The limiting factors were mainly the heat producced by the incandescent lamps and, of course, the cost of the electrical energy. Consequently, in those days the lighting engineer's ideal was to distribute the light in a room as uniformly as possible and with the highest positive efficiency. The idea that finer work could only be done near a lamp was already dead. In those days, when general lighting became for the first time possible, and a high efficiency was essential, reflection from walls and ceiling became important. This led to a complete revolution in interior decoration; light colours and white became quite popular, wall paper and paints appeared on the market which were much lighter and more neutral in colour than those which had been usual at the end of the last century. By the use of evenly distributed light and very light colours and ceilings, all those who were working together in one room had the same and the best possible lighting conditions.

Today, thirty or forty years later, we use lighting levels that are ten times higher and we find that the uniformity considered to be the ideal in 1930 is no longer agreeable at these high levels.

On the contrary, we now seek to employ great variations in lighting levels and brightnesses in one room, similar to the variation in rooms illuminated by daylight, where the level may be 2000 or 3000 lux near the windows, but only 50 or 100 lux near the opposite wall 5 to 7 metres away.

It will be understood that the objects in a room with non-uniform lighting will stand out in much greater relief due to light and shadow than under a uniform lighting that produces scarcely any shadow.

It is also important to know that there is a relationship between the colour of the light and the quantity of it.

This is partly explained by what was said in Chapter VI, namely that where we do not want too much light, as in restaurants, theatres, dancing halls, or in living rooms, the light should be of a warm colour.

In other places where we are more active, as in offices, workshops or sports arenas there should be more light, but light giving a cooler impression (i.e. with less red in it).

Under high lighting levels we should also pay more attention to the surfaces and textures of the things we see. In the first place this helps us to see them in greater detail. It is, for instance, a well known fact that an interior in which a new lighting

Fig. 76. In order to give the reader an idea of the great luminances and the inadmissible contrasts we find on shiny furniture, we show here a desk in varnished wood. The book (the eye task) has a luminance of 100 cd/m^2, the specular reflections show 1000 cd/m^2, a shadow under the book 15 cd/m^2. In general we tolerate contrasts of 1 to 3 between the task (the book) and the environment (the desk).

system has been installed often needs re-decorating and thorough cleaning.
When furnishing a well-lighted room we should choose the appropriate materials. For instance the top of a desk may be made in a fine wood, but in order to avoid uncomfortable specular reflections that wood should not have a glossy finish. These same reflexions would be much less conspicuous and disturbing under less light (Fig. 76).

There are several factors that may cause discomfort when working under daylight. Of course daylight varies greatly with the position of the sun and the condition of the sky (cloudy or clear).

It is impossible to work whilst the sun is shining on our task and we have to use sun blinds.

If we sit around a table at a meeting it is very tiring to have to look at somebody across the table whom we see only as a dark silhouette against a bright window because of the great contrast in brightness (Figs. 77 and 78).

The lighting level diminishes considerably from the window to the opposite wall and there may occur quick fluctuations in the light with the sun and clouds.

In spite of all this we like to be able to follow the rhythm of day and night, of light and dark during our work, so that for most of us, windows, even if they hardly contribute to our ease of seeing, are indispensable as a means of contact with the outer world. They are of course a must for people suffering from claustrophobia.

Now that it has become possible to use high lighting levels in the technically and economically advanced countries of Western Europe and U.S.A., we may enjoy

Fig. 77. Too great a contrast between the bright sky and the dark silhouettes of the people at their desks makes it impossible to see the expression of their faces.

Fig. 78. In this board room, all glare, specular reflections and distracting brightness are avoided. The attention is concentrated on the table and the seats around it.

Fig. 79. In this general office the contact with outdoors is possible by a long horizontal slit of windows. The eye tasks are brightly illuminated by fittings built into the ceiling. The lamps in these fittings burn all day. (Dordrecht)

all the advantages of daylight without the disadvantages of which we have spoken, by combining the daylight with supplementary electric light. It will, however be clear that this supplementary light can be effective only if used in sufficient quantities, such as 500 to 2000 lux.

It might even be said that conditions for easy and agreeable perception during normal work are no longer a question of quick and efficient seeing, but of comfort!

As soon as we realize this we feel that the research and studies in laboratories, that were considered all-important ten years ago, have now lost much of their weight, because the conditions for creating an agreeable visual environment and a comfortable atmosphere are of greater importance than the highest efficiency.

In the richer countries where big and modern lighting installations have become possible, a new practice for the use of these installations has been developed.

If the sky is overcast in the morning and at 10 a.m. the sun breaks through, in modern office buildings, mostly equipped with big windows, the sunshades or venetian blinds are lowered and the artificial light is switched on, even if there would be sufficient daylight left to continue work without the supplementary lighting.

In modern offices and factories, with good air conditioning systems, the drawback to the use of big windows is that the outside climate has too great an influence on the interior temperature.

For visual contact with the outer world, small windows or a narrow horizontal slot (1 metre) in the wall will be sufficient and electric lighting will often be used during the whole working day (Fig. 79).

It is reasonable that in such an establishment the lighting should be at a high level so as not to give the impression of darkness in comparison with the bright outer world.

The use of high level illumination has greatly increased since World War II. This occurred in the first place because the fluorescent lamp appeared on the market in U.S.A. just before the war. It has an efficiency 4 to 5 times that of the incandescent lamp.

It could therefore be said that for the same costs for lamps and energy, five times the amount of light would be available. Others said that one could have the same amount

Fig. 80. A war factory in the United States in 1943. In order to keep the morale of the workers high installations of 2 000 lux (200 ftc) were used day and night for the first time in the history of lighting.

of light at one fifth of the price, which is not quite the same thing! The war made an end of these differences of opinon, when between 1940 and 1945 enormous war industries started in U.K. and U.S.A. Often three shifts had to work day and night in completely blacked-out factories. In order to obtain maximum production the morale of the workers had to be kept as high as possible (Fig. 80).

In the war and munition industries as well as in the automobile and aircraft works, light intensities of 2000 lux were used for 24 hours a day.

After the war, in 1945, news of these installations came to Europe, where these lighting levels were unknown and of course considered excessive.

It must not be forgotten that lighting progress on the Continent had, since 1938, practically come to a standstill and that fluorescent lamps had not yet been used.

In Western Europe and also in North America it was expected that this "excessive" use of light, which might have had its reasons in war time, would soon disappear. In the event, nothing of the kind happened. Once the blessings of abundant light had been experienced, nobody wanted to go without it any more. Gradually this opinion became general in both continents so that the lighting levels for workers jumped from 300 to 500—750—1000 and even 2000 lux — levels that nowadays are considered quite normal.

It is not easy to specify reasonable recommendations for the new lighting practice, and the lighting levels differ in every country for various reasons.

Sitting outdoors in summer, reading in the shade of a tree is very pleasant and comfortable. We may then find 10 000 lux on the pages of our book. This means that recommendations of 1000 to 2000 or even of 5000 lux have only a relative value. They are inspired not only by our well-being but also by economic considerations, or by the maximum admissible emission of heat from the lighting installation, or by the construction of the building.

From all this it is clear that the use of high levels of lighting has become a problem which cannot be solved satisfactorily and quickly by a simple calculation.

Only the experienced lighting expert, willing to work together with other experts and with a keen interest in the well-being of his fellow men, will find ways to arrange working conditions in which we may expect the maximum efficiency of the worker (and not merely the maximum efficiency of his eyes!). The question has now become: how much light do we *want*, and not how much do we *need!*

Fig. 81. Uniformity in lighting generally spells monotony. This example shows how variation in lighting level can be restful and agreeable.

Chapter VIII

WILL WE EVER MASTER THE ART OF MAKING IDEAL SEEING CONDITIONS?

"Indeed his (Land's) works serve to remind us of the dangers of losing phenomena through simplifying situations in order to get neat experiments".

R. L. Gregory — Eye and Brain

During the whole of the industrial era between 1840 and 1940, every worker in industry, in the crafts or in offices did his work under lighting conditions that would now be considered completely insufficient, especially under artificial light. But there were hardly ever serious complaints, because there were almost no possibilities for improving the conditions; and we believe that, in general, not very much harm was done to the health of the workers.

There were, however, a few exceptional situations where the work made excellent lighting conditions indispensable.

Those were artists' and sculptors' studios, diamond cutting workshops and watch factories where the work was done close to windows looking north.

Gradually new light sources of higher efficiency were developed and it also became possible, by employing better heating installations, to use bigger windows. Research was then undertaken into the efficiency of working under the new lighting conditions, in order to find out if this improvement in the lighting was justified economically. Quicker work, less tiredness and the elimination of mistakes should make up for the extra costs of the lighting.

As explained in Chapter VII, it was mainly the war with its intense industrial effort that added a new motive for improvement of the lighting.

It is thus clear that the decision about the size of a lighting installation depends on many different factors, which in their turn often vary with time. It is also clear that recommendations cannot be accepted internationally, since economics, the climate, and the traditions of the different countries, sometimes even of different provinces make local decisions necessary. Today in many countries we find an appreciation of the advantages of a good lighting level during working hours and since the higher costs are no longer a serious matter in the economically-strong countries, we find there hardly a need for recommendations as to minimum lighting levels.

It should not be forgotten, of course, that the recommended lighting level is by no means the last word to be said about the quality of an installation.

The requirement as to the appropriate colour of the light, the correct use of shadows, high lights and reflections, the absence of glare etc.... are all important factors.

But we find additional factors of importance which influence the quality of the high level installations, concerning which it seems impossible to lay down hard and fast definitions, because emotional evaluations arising from human reactions cannot be calculated with any degree of precision.

— When we install high lighting level illuminations we find that the nearer we approach daylight-levels, the greater is the desire to achieve the colour of daylight also.
— In a large room where many people work together, and where a uniform lighting level is inevitable, this uniformity leads to monotony, for which some form of correction is desirable. This may be done by decorating walls and columns lying within the normal visual field of the workers, with striking rather saturated colours.
— In smaller rooms this monotony can be avoided by arranging that areas near the windows where most of the work is being done, receive the full lighting intensity, leaving a strip opposite the window much darker. This arrangement is analogous to work by daylight where intensities in the region of 2000 lux near the windows may exist but only one tenth this value at points 4 to 5 metres away from them.
— One often feels, after a long period of concentrated looking, the need to rest the eyes by looking without concentration, perhaps even without seeing, at a dark corner in the room. This may explain the desire to avoid uniformity in the lighting (Fig. 81).

The aspect of the installation itself creates another problem due to the fact that the large number of lamps and fittings required for high intensity illumination must in general be mounted on or in the ceiling where they are usually not only conspicuous but do not enhance the appearance of the room (Fig. 82).

Hitherto the general practice has been simply to choose a lighting level, then to calculate the necessary luminous flux, select the type and number of lamps and fittings to be used and to distribute these evenly over the ceiling. This method rarely

82

Fig. 83. A modern installation in an office with a high lighting level, but the distribution of the fittings in the ceiling is monotonous and irritating because of the excessive brightness contrasts.

constitutes an embellishment of the interior (Fig. 83). If the lighting expert wishes to seek a better solution, he will have to visualize the room as it will be seen by the worker. He must avoid making the ceiling an excessively dominant bright element in the room; he should also avoid glare, if that is possible. For that reason one should also avoid reflexions on the task and its environments. These reflexions are a serious drawback of all the modern lighting systems we know. It has been tried to find a solution for this problem by using polarised light, sometimes it is possible to choose mat and more diffusing materials for the eye task (e.g. mat paper) and its environment (the desk top) but the light itself always presents this grave disadvantage (Fig. 84).

We have to try and find constructions and shapes that make the ceilings and the lighting system as discreet as possible. Sometimes it is possible to mask the light sources completely by screens or girders, if those working in that room are all looking in the same direction (Fig. 85).

In order to find the best possible solutions it will be necessary to establish a close co-operation between the lighting expert, the architect, the decorator or a colour specialist. Their main objective must always be the well-being of the worker. In this way it will be possible to produce pilot installations which will have a great influence on the further development of good and up to date working conditions (Fig. 86).

Once again we want to stress the necessity of composing the complete visual field of the worker, so that special arrangements, colours and textures of every visible surface play their part well. The floor, the furniture, the walls, the machines, the desk tops and even the paper that is used are part of the total image which means that even the purchase department of the firm should know about this and perhaps sometimes have a representative in the team (Fig. 87).

Good lighting by daylight alone is very rarely achieved. In the first place the constant variation of the direction of the light; sunlight, clouds and seasons make it impossible to obtain the light that is needed, and to ensure that it is sufficiently constant and of more than minimal intensity.

There are other reasons why daylight is not very suitable for work. The architect is rarely in a position to arrange ideal orientation of the windows. Most of the higher buildings have their windows on all four elevations. The use of much glass presents great difficulties in keeping out the sun and in ensuring a constant temperature, since the low thermal capacity of the outer walls causes the temperature to vary rapidly with the weather and the wind.

In rooms which derive their illumination from windows, a lighting level fit for normal work occurs only within distances of 3 to 4 metres from the windows. This means that it is impossible to use the whole floor surface efficiently without supplementary artificial light.

Against all these objections we must certainly set the invaluable advantages of visual contact with the outer world. This is without doubt the reason why windowless offices and factories are used only in extreme cases — usually for reasons of constant humidity and temperature as in textile factories, printing works and tobacco factories.

Daylight is used chiefly in the home. Here the inconveniences of direct sunshine, of good light near the window only, and inconvenient orientation of the windows are easily corrected by the resident. He is much more mobile here than when at work, he can choose the place where he sits and can use curtains or sun blinds how and when he likes. This does not mean that the use of daylight in the home has not often become more difficult in modern houses than in older ones. The reasons have been explained in Chapter III.

There have been new developments since 1945.

As has already been pointed out it has become possible with the new light sources, to obtain lighting levels with artificial sources that greatly surpass the levels of daylight indoors.

The first consequence was that rooms originally illuminated by daylight in which there was only sufficient light at distances up to 3 m from the windows, could be equipped with supplementary artificial light opposite the windows so that the whole area of deep rooms could be used as a working space, and points 6 to 10 m distant from the windows become equivalent to places near the windows.

Fig. 88. The v
are being drav
The blinds of
are still open
the offices are

The influence of daylight on working interiors is being reduced to a minimum. Windows are considered only as the means to maintain contact with the outside world. If the windows are large (curtain walls) they are reduced in size by curtains, venetian blinds and sunshades (Fig. 88).

We have seen excellent installations in Cleveland and New York. We saw very large offices (for 100 to 200 employees) in which the windows were shut off with venetian blinds. There were complete air-conditioning systems, some with special cooling ventilation of the lighting fittings. The light was almost completely screened off sideways; looking up from the desk there was an interesting display of gay and bright colours on walls and columns. Our only comment would be that the lack of contact with the outside was a disadvantage, but this seemed to be of little concern to the employees. There were also the drawbacks which we find in practically all high level installations, namely that there were reflections in the desks, and that the papers on them were too bright. These disadvantages are very difficult to avoid. We believe that at present the remedy can be found only in the judicious choice of the colours, surfaces and textures of desks, papers and typewriters.

We saw several good solutions at the University in Cleveland. Professor Putman had an office there. It had small windows with venetian blinds that screened the windows up to 1.5 m from the floor. Sitting down one could just look out. There was no glare from the sun or from a bright sky. The installation of fluorescent lamps, blending in colour with the daylight, produced an illumination of 2000 lux (200 foot candles) on the desk. This provided an excellent working atmosphere.

Those who, like the author, think that this is the road we shall have to follow, will understand that most buildings, even the recent ones, are still far from being ideally designed and equipped. This means that the very best working conditions can be realized only with the use of what we still consider very expensive installations.

If we want a level of 750 lux (75 foot candles) in a room, a good ventilation system is essential, and for levels between 1000 to 2000 lux the air-conditioning system must be combined with a ventilation system for cooling lamps, ballasts and fittings, not only in order to reduce the heat in the room but also to ensure higher efficiency of the light sources.

It may be said that modern lighting systems had their origin in the installations used to supplement daylight. Gradually these systems, whose size was originally limited by economic considerations, changed into installations in which daylight was no longer a factor and in which optimum working conditions have now become the only criterion. To what extent this way of thinking leads to practical applications depends greatly on the prosperity of the country and on its technical development.

We have already insisted several times that we can create good working conditions only if we consider all the factors influencing these conditions, i.e. factors concerning light (its colour, direction, quantity), space, colour, sound, acoustics, climate and contact with the outside world.

The man who has to develop these conditions, or better still the team of men who have to deal with them, must be familiar with them and be able to handle them.

How will they obtain this knowledge?
The only way is to seek personal experience in existing offices, workshops and industries and by learning how these factors can influence one another, for knowledge about this mutual influence can also be obtained only by experience.
We will give a few examples:

- Work in a room which is too hot becomes more acceptable if the room is illuminated by ample cool (that is blueish) light.
- Monotonous work becomes more bearable in a lively environment with much colour, also by providing music during the working period.
- If intense concentration on the task is necessary, as in the case of very fine work, it can be encouraged by good composition of the visual field, (see Chapter V).
- Brainwork rarely calls for special environmental conditions for the worker, so that normal requirements can be met by neutral colours and small contrasts in the visual field, no movement, silence and not too much light.

How can the members of such a team learn about all those factors and how to handle them?
It is our experience that a long period of joint study, research and development, during which the team visits and judges the different kinds of work, the places where that work is done and the lighting installations, is the only answer. This has to be done, keeping constantly in mind the attitude of the worker. It has to be done collectively because only then one can find the points where the different groups of factors have a relationship.
Each examination and judgement must be made in daylight and in the evening; there must be contact with the workers; their eye tasks must be studied, and if possible the observers should actually undertake the work for some time.
After these visits the members of the team should discuss their impressions and try to come to unanimous conclusions.
Such analyses lead in the first place to a very useful self-education and instruction of the observers. As we said, the interest in the well-being of the workers, our fellow men, should always be the target of the whole study.
In 1955 the *Congrès International de l'Eclairage* embarked on this work and study in Committee E 3113. The most important work this committee on "Agreeable luminous environment" has done was the preparation of questionnaires to be used for analysing existing interiors — a work which has covered many years, and from which the members of the committee have learned a great deal.
At the end of this book the text of such a questionnaire has been reproduced because we believe it to be an exceedingly interesting and instructive document which proves, when being used, that the lighting art can only lead to better working conditions in co-operation with many other specialists (see page 132).

89

Chapter IX

THE EYE IS BLIND TO WHAT THE MIND DOES NOT SEE

Good visual environment means:

a. An eye task that can be seen easily (light, colour, shadows, reflections, no obstructions).

b. Surroundings of the eye task and background are conducive to ease of perception.

c. An environment in harmony with the visual concentration in relation to the character of the work.

When considering these three factors influencing our visual world, we must never forget that all the impressions that come to us are finally received, analysed and integrated by the human mind. This means that our aim really is to try to make the average observer's reactions to what he sees as perfect as possible.

We will give a few examples of situations in which the environment has an obvious relation to the kind of work to be done.

— If one has to do mental work, internal concentration is necessary.
  Visual impressions should be vague and unobtrusive, which implies that the total image within the visual field should be as neutral as possible, for instance without saturated or strong colours, high brightnesses or large contrasts.
— If one has a controlling function, which implies long and constant visual attention without actually seeing much, the visual field should be quiet and not distracting (Fig. 89).
— In a meeting, sitting around a table, one has to see three different tasks: papers on the table, the expression on the face of a speaker, or just all the persons around the table during a general discussion. Then the table is a visual centre and should be treated accordingly (attractive wood, matt, smooth surface, good light) (see Fig. 78).
— When working with small components and precision instruments the eyes have to be used intensively, fixing them on one small part of the visual field. That part then should be emphasised by high intensity lighting that makes the brightness pattern of the rest of the visual field uninteresting (Fig. 90).
— For very regular and almost automatic work the eyes have little to do and the visual field may be rendered more attractive by colour, brightness and design. Women can knit for minutes without looking at their work. Girls packing small boxes or bottles in crates for hours at a stretch can do so without looking; only when taking a new crate will they glance at it.

90

For such work the background can play an important role.

— This is also the case with children sitting in a classroom.

The wall towards which they are looking should have a composition (see Chapter V) that holds their attention!

We saw in Chapter V that seeing is an activity which is interrupted and redirected hundreds of times per minute. We *look* longer and more often the more our eyes are attracted, but we *see* only those things with which our mind is occupied.

During a walk, the hunter will see the game, the biologist the plants, the child the flowers, the painter the landscape and the man who never learned to walk will see nothing and be bored.

Men looking at a lovely girl will see her face or her figure, women looking at the same girl may see only her dress.

A man can drive for half an hour along a very fine road with his wife sitting next to him. At the end of the road both may say "that was a fine road" and they will mean two completely different things: the man did not see the landscape, the woman did not see the road!

Also we often see the things we like or expect to see and conversely we do not see the things we hate to see.

Somebody who likes sensation will see an accident on the road as a serious one, whilst an unemotional type will try to see it as being of minimum gravity. Somebody with a logical mind will see what he imagines to be possible or what can be explained. He has a tendency to ignore improbable happenings.

We are struck by the unusual and are then inclined to call beautiful what we would not find beautiful if we saw it frequently.

We no longer see the tree we pass every day until something unusual is happening to it. In winter it is a bare silhouette of dark branches against a pale sky. It suddenly arouses our interest when we see the same silhouette in white frost against a blue sky or dark clouds. We can obtain a similar effect if we look at a photographic negative of the same tree. The Japanese like to see trees and plants as silhouettes: they take their cameras out when there is a mist (see Figs. 91, 92, 93 and 94).

92

93

Frequently we are inclined to consider the unusual as beautiful. Only artists can see everyday things as interesting objects, because their mind is always open to what they see.

Ugly things, annoying things or things we should like to forget are easily overlooked. (He does not see the mess in his room; she turns a blind eye to him).

Summarizing these points, we might distinguish four different ways of looking:

1. Looking without seeing (staring, looking out of a window during a conversation).
2. Looking automatically, during which we only see one certain object (when driving a car we look uninterestedly at the road till we see a car coming our way or till we have to look with interest at a traffic sign. A man working in the control room of a power station is looking without interest till a red lamp attracts his attention).
3. Looking almost uninterruptedly at something that interests us (exacting work, reading and also watching ballet or cinema films).
4. Looking around with only now and then an intense observation (operating an automatic press or a loom, but also taking a walk or when shopping).

Concerning paragraph 2 above, we would remark that automatic looking depends greatly on our attention. One can have a conversation while working, or listen to the radio whilst driving. The more interesting or heated the conversation, the more catchy the tune, the more our attention is distracted and the less we actually see.

There even comes a moment when our mind is so occupied by what we hear we no longer see anything and accidents are thereby caused.

All these considerations convince us that the designer of working conditions, including the lighting, must be thoroughly acquainted with the character of the work in question. He should even have to do that work himself for some time. It is also necessary that the observers of whom we spoke in Chapter VIII should be trained observers knowing the work well and its effect on the mind of the worker.

It seems strange that hitherto little attention has been paid to this study of the

93

94

character of the work, because this means that the planning of lighting installations has seldom been done with the human factor in mind.

And this certainly must be done!

At first sight this looks like involving an enormous amount of work, but the task may be less extensive than might be anticipated.

In the first place there are groups of activities that demand similar working conditions. These conditions, once analysed and laid down, will remain valid for a long time. Secondly they will have to be changed only when the working methods are changing, or when new lighting methods or light sources are developed.

In the third place we believe that this research could be done internationally, i.e. that experiments made in one country could also be accepted in other countries as the bases on which lighting practice should be built up, provided those countries do not have widely differing climates and living conditions. We are thinking here of countries and continents lying between 40° and 60° latitude in Europe, North America and Australia.

Chapter X

DEVELOPMENT OF LIGHTING TECHNIQUE

Most of the advances in science and industry during the past hundred years have been improvements in technique leading to lower production costs. This is understandable, since profit was the principal motive for these developments. Humanitarian considerations came a very poor second, if indeed they played any part at all.

Activities like weaving, the casting of metals, the tanning of leather and so many others, had to be carried out under disgraceful circumstances right up to the Second World War. Chemical industries pollute air and water and it is only afterwards — when the damage becomes unbearable — that measures have been taken to reduce the harmful effects. Insecticides and chemical fertilisers disturb the balance of nature and are exterminating more and more species of insects, birds and fish. The automobile industry produces more and more cars, and only after half a century is some attention being paid to safety and to the complete combustion of petrol. The building trade is looking for even more efficient and rapid methods, but it is hardly concerned with the way of life and the needs of those who will have to spend the next 25 years in the houses it builds. The hundreds of thousands of dwellings that are being built now will be a terrible obstacle to better living in the near future.

And indeed, it is difficult for a research worker or a technician to pay attention to the human aspects of his technical problems. His employer would probably object if he did so, fearing that it might influence the progress of his work. However, in this chapter we would like to spend a little time considering what "creative lighting" really means for the people who are to use it.

As long as man has been building homes for himself and for his family (probably the past ten thousand years at least), light has played an integral role in the conception of a building, as regards both visibility inside and an attractive appearance outside. In this book, we speak of "creative lighting" since we think that an activity which can benefit man so much and can engender so much well-being, comfort and beauty, deserves the epithet "creative". It is no longer necessary to persuade our readers that electric light can be used just as creatively as daylight (or sometimes even more so, since we can vary the siting of our light sources and the direction, colour and intensity of the light within very wide limits). Artificial light in the form of flames from oil lamps or candles was often very beautiful, but very expensive, so that it was mainly used in churches, halls and theatres; and even there its use was subject to strict limitations because many flames produce not only much light but also a potentially dangerous amount of heat.

The first electric light was lit in 1801 by Davy in England, when he invented the arc lamp. However, it was not until 1879 that Thomas Edison made the first practical incandescent lamp; and before this could be used in large numbers, all the problems concerned with the production of electricity, on a large scale, with wiring and switchgear had to be solved. The first installations were expensive, so they were used in spectacular projects, like the lighting of the steamship Columbia in 1880 (115 lamps), the World Exhibition in Paris, and the Crystal Palace in London.

Arc lamps lit many squares, theatres and ballrooms at the beginning of this century. Before 1920, incandescent lamps were small, expensive and inefficient. Lamps of at most 10 to 12 candles were available for local lighting during work.

Between 1920 and 1925, incandescent lamps became more efficient and reliable (tungsten filaments, gas-filled bulbs, coiled and coiledcoil filaments) with the result that gradually electric lighting could be used for the general lighting of working spaces. It is understandable that in those days the manufacturers of incandescent lamps concentrated on questions of duration, efficiency, and standardization of the shapes and sizes of the lamps.

The *Commission Internationale de l'Eclairage* (C.I.E.) was founded in 1904. Its first activities were the study of photometry, the definition of lighting units, investigation of the life and efficiency of lamps, and standardization.

When large lighting installations came into general use, the need for uniform illumination began to make itself felt. Reflexion from walls and ceiling, shadows and similar effects now had to be taken into account in the design of such installations. New methods of calculating lighting levels were developed, and the amount of computation involved was reduced by the tabulation of certain results (Harrison &

Anderson[1]). But the reason why these calculations were introduced was that even lighting over the whole surface of a workshop allowed the whole floor to be efficiently used for work, and not that uniform lighting was agreeable to the worker. Man is accustomed to the great differences in lighting between day and night, sun and shade, indoors and outdoors, and he does not greatly appreciate uniformity.

Incandescent lamps became bigger and bigger, and the filaments became brighter and brighter. People started complaining about glare. This could be partly remedied by making the lamp bulbs mat or of opaline glass. For bigger lamps (100-200 watt), large diffusing fittings of opaline glass had to be used. Once again, the manufacturer did no more than just deal with complaints from users.

The above-mentioned calculations showed that the reflection of light from walls and the ceiling has a great influence on the efficiency of the installation. Lighter colours and white thus became more and more popular for walls and ceilings in this period (1925-1930), and sales of light-coloured wallpapers rocketed. However, electric light was still considered costly. Industry therefore did its best to make installations, lamps and electrical energy cheaper so as to enlarge the market for its products.

A number of factors that play a role in the application of electric lighting were studied at this stage: if it could be proved that more and better lighting raised the productivity of the worker and reduced human errors and fatigue, that would be a very good argument for the use of more lamps and electricity. Lukiesh tried to measure fatigue by the variations in blood pressure and the frequency of movement of the eyes[2]. Other workers (e.g. Weston)[3] investigated the performance of two equivalent working teams carrying out the same task under different lighting conditions. Laboratory studies of glare were carried out, with test subjects looking at light sources of various brightness against different backgrounds (Hopkinson[4], Guth[5]). However, here again this research was inspired by the interests of the industry, and the results were incomplete since the tests were carried out in surroundings that were very different from the places where the fittings were to be used (e.g. no colours).

Weston, an officer of the British Ministry of Health, studied the speed of perception in a visual task (observation of Landolt rings) under various lighting conditions. In the U.S.A., the "optoscope" was invented, and used to allow a great number of observers to choose for themselves the lighting level they prefered for a given reading task. It appeared from these experiments that the need for light increases considerably with the age of the observer. In order to investigate this point further, Fortuin[6] examined a great number of persons, and found that human visual acuity diminishes from the twelfth year onwards, so that a middle-aged person needs 10 to 15 times the amount of light (for a given task) that someone of 18 needs.

American studies about glare (Guth) led to the introduction of a semiquantitative glare scale running from unbearable via more or less uncomfortable to bearable and just perceptible. Past this last point, a no-glare line is called the "borderline comfort-discomfort" (B.C.D.). This quite wrongly suggests that the absence of discomfort implies comfort!

In 1951, at the C.I.E. congress in Stockholm, the author spoke of the desirability of using the worker's overall impression of the quality of a lighting installation as one of the main criteria. The man who has to work in the room is the only one who can really judge what the working conditions are like!

It is obvious that the quality of a lighting installation does not only depend on the intensity of the light; the shape and size of the illuminated space, the colours and textures of all the visible surfaces and the brightness pattern on them play just as important a role.

Our knowledge about visual comfort is still rather incomplete, and research in this field is difficult because of the large number of factors involved.

In 1955, Balder[7] tried to determine the optimum brightness for visual tasks in normal office work. However, he did so in the laboratory, in an environment without colour, shadows or visible light sources. In 1962, Bodmann[8] carried out similar experiments in more realistic interiors, with a decoration scheme involving various colours and textures and a representative lighting system. Barthès[9] reported on similar efforts at the C.I.E. congress in 1963; he worked with small model rooms in which he could vary the colours of the walls. In the U.S.A., Helson[10] has carried out experiments with a number of test subjects to determine preferred colour renderings and colour harmonies in interiors. Many of these experiments at last represent an attempt to discover that the person who is going to use a lighting installation really needs (wants).

In the meantime, very important technical progress had been made. Fluorescent lamps, developed and improved after the Second World War, made it (technically and economically) possible to use considerably more and better light, while the development of improved fluorescent powders for these lamps gave much better colour rendering for different visual tasks, and the great variety of colours allowed a harmony with the different room atmospheres. Better ventilation and air conditioning, designed together with the lighting system, can now deal with excessive heat from the lamps.

In 1954, as a result of a paper read by the author at the C.I.E. congress in Zürich, a commission of experts (E 3113) was set up to study all the factors that play a part in creating an "agreeable luminous environment".

Two possible lines of research were considered. In the first place one could install two comparable spaces, decorated and lighted in different ways, in which observers could work for a certain period and then express an opinion about the advantages and disadvantages of the two. This proved to be a slow, difficult and costly method which yielded very few results, since the various observers, being different persons, had very different motives for their preference for the one or the other environment.

Another possibility seemed to be the analysis of existing interiors which were generally recognized to be good or bad. If it proved possible to find the factors which make one working interior preferable to the other, this might lead to some guide lines for new designs. For this purpose, questionnaires were composed dealing with the main factors that define the quality of interior lighting, such as: the visibility of the object in question, the colours and brightness within the field of vision, the amount of room glare, uncomfortable contrasts, etc. (See Chapter 1 and the appendix at the back of this book).

In practice, this method proved tedious and time-consuming; and the results, representing a combination of individual opinions, were often vague and erratic, and sometimes influenced by completely unexpected factors. However, the work of composing the questionnaire was most instructive for the experts, who came to realize that good lighting is an art which requires great talent, considerable experience and much interest in our fellow men.

The expert commitee E 3113 was active between 1954 and 1967; a new committee (E. 16) was installed at the Washington congress in 1968. The committee included the total environment — i.e. lighting, climate and acoustic conditions — in its studies. Activities directed towards control of the total environment are often referred to as "human engineering" in the literature. However, we would like to argue that this term is not very suitable. If men are to work efficiently in a given space, a large number of requirements must be met.

As regards the (micro) climate:

the right temperature
regular renewal of the air
safe radiation level
controlled humidity

As regards lighting:

sufficient light for the task in question
good colour rendering
correct direction of the light (shadows)
absence of glare.

and as regards the acoustic conditions:

not too much noise
not too much silence
the right amount of reverberation.

Realizing these conditions is indeed an engineering job, which is normally regarded in negative terms by the specialists concerned:

not too cold or too warm
no stuffy air
no radiation hazard

not too little light
no bad colour rendering
no indistinctness
no glare
not too much noise
not too much silence
not too much reverberation.

These negative requirements can largely be met with the aid of the instruments and formulae available now, but insufficient attention is still being paid to the factors that bear on the positive wellbeing of the individual, completely ignoring the psychological factors.

The engineer certainly knows the beautiful sparkle of sunshine in the water, the moving pattern of shadow under a tree, the festive glitter of crystal chandeliers and jewels in a ballroom. He knows the beauties of landscape, ever changing in aspect from sunrise to sunset and the thrill of a brilliant snow-topped peak seen from a green valley; he has surely experienced the charms of music and words, and he knows how just humming a tune can break the monotony of a long working period.

The air-conditioning expert knows that with all his knowledge he will never be able to produce the delightful sensation of sitting on top of a sand-dune, looking at the sea in a gentle breeze on a hot afternoon.

We are not suggesting that the worker should be put on a sand-dune to make him happy, but we would stress that the time has come for the designer of industrial premises to consider not only how the worker's discomfort can be alleviated, but also how his creative abilities can be released to the full.

We should therefore like to introduce the term "human design" for this activity. In order to create the ideal environments, a new kind of designer is needed, a designer who will base his work on human desires and needs, and who will also consider the positive factors creating comfort and well-being. He should be the man to find all the experts with whom to co-operate in order to realize the best total environment.

At the moment, it would be difficult to find a single "human designer" in this sense,

but it must be possible to train them. At this moment there is no training available for such specialists, but the institutes of technology where the architects, electrical, acoustical and airconditioning engineers are trained might be the place where a team (including a psychologist) could take such an educational task in hand.
We want to emphasise again that the whole development of the technique of lighting, up to this day, has been inspired by the 19th century conceptions of society, ignoring completely the well-being of the worker and letting commercial interests dominate all research.
It is only during the last two decades that the social evolution in Western countries has led a small number of lighting experts to consider the well-being of the individual during his daily work, the main *leitmotif* in their work.

REFERENCES

1. Harrison & Anderson, Coefficients of Utilization, *Trans. ill. Eng. Doc. N.Y 15*, 1920.
2. Lukiesh, M. and Muss, F. K. "The Science of Seeing" 1937; "Light, Vision & Seeing", 1944, van Nostrand, New York; "Visibility— its measurement and significance in seeing" *J. Franklin Inst. 220*, 1935, 431.
3. Weston, H. C., "The relation between illumination and visual efficiency— the effect of brightness contrast", Report No. 87 Industrial Health Research Board, Medical Research Council H. M. Stationery Office London, 1945.
4. Hopkinson, R. G., "Evaluating of glare", *Illum. Engng, 52*, 1957, 247.
5. Guth, S. K., "A method for the evaluation of discomfort glare", *Illum. Engng, 58*, 1963, 351.
6. Fortuin, G. M., "Visual power and visibility", *Philips Res. Rep. 6*, 1957, 251 and 347.
7. Balder, J. J. "Erwünschte Leuchtdichten in Büroräumen", *Lichttechnik Berl. 9*, 1957, 455.
8. Bodmann, H. W., "Illumination levels and visual performance", Int. Lighting Rev. *13*, 1962, 41.
9. Barthès, M. E., "Etude expérimental de Mons. Farnay et Barthès pour la mise en point d'une méthode de calcul du point de couleur de la lumière reçue par le plan utile dans un local à parois colorées", *Bull. Soc. Fr.* Electns *Serie* 7, 1957, sub.
   "Essay de synthèse des connaissances sur la physique et la vision des couleurs en vue de la prédétermination de l'ambiance colorée d'un local", P. 63.11 C.I.E. Proceedings 1963, Vienna.
10. Helson, H., "Pleasantness of object colours as a function of source and background colours", *West. Architect Engnr*, June 1961.

The Appraisal of Office installations

INTRODUCTION

The C.I.E. Committee E.3.1.1.3 tries to find methods which may lead to a
ledge of what makes the visual environment in working interiors agree
The first years of study made it clear that his subject is so big that in
able to get workable results the first studies should be very limited in s
first phase the committee wants to study normal offices of medium size, t
meant for at least 4 persons. The offices to be chosen for this questionna
among the best you can find in your country since it would be impossib
conditions for good and agreeable seeing in offices that must be considere
respect.
We think that may be there is a distinction between the conditions for agr
environment in offices of a very high lighting level between 600 and 2
offices of 500 lux or less. It would be interesting for that reason to have
examples of both categories.
The questionnaire which will be used for this study consists of 4 parts. T
parts are meant to obtain subjective information on the installation an
the fourth part to obtain the objective data.

Those four parts are:

I. the observer's first general impressions of the office

II. the observer's judgment of the luminous environment when doing t
office is designed for

III. the comments of the operators working in the office

IV. the objective description of the office and its luminous environment

For each installation are needed:

3 copies of the parts I and II (one for each observer)
1 copy of page 5 (report on discussion)
3 copies of part III (one for each operator to be interview
1 copy of part IV

The observers are requested first to fill in parts I and II independently of
and then to compare their answers trying to find out the reasons for diffe
answers. These reasons can be reported on page 5.
Since it is difficult to obtain the right answers from an operator who is nc
observer, considerable tact is required from the observer not to influence the
answer. For that reason the team of 3 observers may decide to appoint a fo
with psychological training who would ask the questions from the opera
the observers are filling in the first parts of the questionnaire. This inter
take place in another room.

EPILOGUE

"…. it is up to our engineers and our architects to reveal clearly that the aim of the technical progress is to help the human being in his quest for his autonomy and the fullness of his being.

Jean Fourastié

This book has been written for various reasons.

— In the first place in order to put on record the experience gained in 45 years of looking conscientiously and critically, trying to analyse why the things we see look good or bad.
— Secondly we should like, by this book, to give the technique and the art of lighting the standing and the place it deserves among so many other new technical sciences.
— Finally, we have tried to prove that the human factor is of prime importance when we try to find the optimum visual conditions. This human factor has, hitherto, been sadly neglected, as it has been in many other technical sciences which are too often considered as of a purely physical nature.

It is disappointing to observe that during the sixty years that electric light has so greatly influenced the life of the populations of all developed countries, that influence has been mainly limited to technical matters and economical problems, industrial and commercial.

Other senses such as hearing, smell and taste play a much smaller part in the impressions we get from our environment, but they have had much more attention than seeing, as evidenced by progress in acoustics, air-conditioning and ventilation, and an ever-increasing interest in food: its vital components, international recipes, its influence on our health, and dietary in general. Thorough study of visual perception has only just started!

Appendix (see chapter VIII)

This questionnaire (page 133-138) was made in c
and to discover the factors that have an influ
environment in working interiors'.

PART I

The observers first general impressions of the office.

As you know the first impressions are not always the same as those one gets after some time of acclimatisation and after choosing a definite working position. At this moment we only want to know your first impressions which you should give without extensive considerations about the correctness of your answers. We invite you to gather these impressions by walking around the office, looking in different directions and making a general survey. Then sit down in the place of one of the operators and answer the following questions.

1. Try to describe in a few lines as completely as possible the first general impressions of the room you entered:

The room is rather dull and untidy, the outlook from the windows is uninteresting. The work done here seems to be rather monotonous so we would have preferred more gay surroundings.

2. Can you qualify your first impressions as being:

unfavourable

indifferent —— definitely indifferent

good

excellent

3. What do you think is **strikingly** good in this room?

the disadvantage of this room is that there is nothing strikingly good.

4. What do you think is **strikingly** bad in this room?

the monotony and the lack of colour.

PART II

The observer's judgment while doing work similar to the kind of work the office is designed for.

After you have been sitting some time in the operator's place doing similar work the office, is designed for, you will be able to give a more founded judgment about the working and seeing conditions in the office. Please formulate this judgment by answering the following series of questions.

In this column merely put a mark behind the word that expresses your opinion best	In this column, please write down the reasons why you have marked the word you did. You may also give in this column further comments or suggestions for improvements
1. Does the general lay-our of the room appear: untidy satifactory X effective Think of: arrangement of furniture and other equipment; dimensions of the room, partitions; etc.	Why? dimensions are good, lighting fittings too obvious furniture is placed without much system or order.
2. Is the appearance of the furniture: unattractive X acceptable attractive	Why? design of furniture and material and colour is rather bad.
3. What do you think of the whole colour scheme? unattractive X indifferent attractive Think of: colours of walls, floor, ceiling, furniture, doors, accent colours, etc. and the colour harmony.	Why? hardly any colour, mostly grays and dull metal, ceiling white floor brown lino.
4. Can you see the essential parts of the operator's task: with difficulty not so well rather well X easily	Why? —
5. Are the surfaces and objects in the room revealed: with difficulty not so well rather well X easily Can you easily and quickly take in: 1) the size and shape of the room 2) your own position in relation to walls, ceiling etc. 3) the character of the materials and surfaces To form your opinion you may walk around making sure that objects are what you took them to be, and look e.g. at your own hand.	Why? —
6. Is the visual environment while concentrating on the work: distracting of no special value X assisting concentration	Why? —

7. When looking up for relaxation and moving your eyes about, do you easily find a visual rest centre? yes no X If you do not feel the need for such a center, neglect this question	Why? there is nothing that might interest, not even the outlook from the window.
8. What in your opinion is the quality of the light in the room: unfavourable indifferent X good excellent The light can be: dim or bright: harsh or soft; cold or warm; variegated or uniform	Why? too uniform, not enough in daytime
9. Do you find the appearance of the lighting installation unfavourable X indifferent good excellent	Why? ugly black and white, metal fittings, too much contrast between outside and inside of reflector.
10. How is the glare of the lighting instalation? intolerable uncomfortable just acceptable X unnoticeable	Why? too much contrast between fittings and background.
11. Are specular reflections in the room? dazzling irritating X distracting of no importance contributing to the vividness of the environment	Why? The dark shiny tables and desks give irritating reflections.
12. How is the colourrendering in the room? poor fairly good X good (you may pay special attention to the complexion of other persons in the room)	Why? the faces of other persons in the room have the right colour.
13. What do you think of the brightness distribution? disagreeable X neutral agreeable	Why? the pattern of the fittings on the ceiling is irregular and too obvious the white walls should have a darker and more friendly colour.
14. You have now examined the nature of this room and its contents, the seeing conditions and the lighting. If you now sum up all these details how would you then qualify the total impression of this office? disagreeable } X indifferent } good excellent	between disagreeable and indifferent!
15. It may well be that after the close examination of all the details in this room and after having been in it for a long time you feel that there are differences between your first general impression and the judgment you have qualified sub. 14. Please try to describe these differences here:	hardly any difference.
16. If there are considerable differences in the visual field of different operators in the room, please note it here, indicating the cause of these differences.	no, everybody is, during work, looking in the same direction.

PART III

Comments of the operators working in the office

Since the operator in general is not a trained observer or a lighting expert it is impossible to let him fill in part II without any further assistance. Moreover, many of those questions cannot be explained in a few words in such a way that he could give a valuable answer. For these reasons it has been decided to ask only a selection of the questions from part II and to do this in an interview. The interviewer must be careful not to lead the operator's answer. While interviewing the operators it should be borne in mind that their opinions may be easily influenced by factors which have nothing to do with the visual environment itself: e.g. the answer may be rather negative if the operator is not content with his work.
The interviewer must also be careful not to point out deficiencies in the office of which the operator has not been aware. The operator must know that he is free to give his answers and that they will not be reported to his employer. The interviewer should try to find out whether the answers would be different in case daylight plays a part in the lighting conditions and report on these differences.

In this column merely put a mark behind the word that expresses the operators's opinion best.	In this column please write down the reasons the operator has given for his grading and eventually differences between day and night.
1. Can you see the essential parts of your task: with difficulty not so well rather well X easily	daylight hardly plays a role in the seeing of the task The electric light is sufficient but hardly creates an animating atmosphere.
2. Is the visual environment while concentrating on the work: distracting of no special value X assisting concentration	—
3. When looking up for relaxation and moving your eyes around, do you easily find a visual rest centre? yes no X If the operator does not feel the need for such a centre, please neglect this question.	no, hardly anything of interest to be seen
4. How is the glare of the lighting installation? intolerable uncomfortable just acceptable X unnoticeable	—
5. Are specular reflections in the room? dazzling irritating X distracting of no importance contributing to the vividness of the environment	yes, the desk are too glossy.
6. How is the colourrendering in the room? poor fairly good X good (the operator may pay special attention to the complexion of other persons in the room)	there is very little colour in the room, but the faces of collegues seem natural in colour.
7. What is the operator's total impression of his office? unfavourable indifferent X good excellent	I never found the room very interesting or inspiring.

Report on the discuissons about the difference in answers

After the observers have filled in parts I and II and then compare the answers, they will probably note differences in the answers from different observers to the same question. There can be various reasons for such differences: e.g. each question covers the effects of more than one physical property of the office and will be weighted differently by different observers; each observer reacts different to differently physical properties, etc.
The reasons for the differences should be found out in a discussion between the observers. Please report on this page on this discussion.

One observer thought the lighting to be good, two found it to be just sufficient, but not ample, specially in day time. Everybody agreed that there was very little colour or attraction in the room and that it was rather untidy and a bit dirty. If the lighting had been arranged more on one (the left) side of the desks, so as to be less uniform and more concentrated on the desks, the room could be much improved Some colour on the walls and perhaps natural wood for the desktops would also have improved the atmosphere in the room.

ACKNOWLEDGEMENT

Photographs	1, 2, 5, 13, 18, 19: Keuzenkamp, Delft
Photographs	3, 4, 16, 23, 24, 31, 32, 33, 34, 35, 36, 41: Lucien Hervé, Paris
Photographs	6, 12, 14, 15, 39, 46, 51B, 74, 76, 77, 78, 79, 80, 81, 82, 83, 85, 89: International Lighting Review
Photographs	8, 17, 30: National Monuments Record, Great Britain
Photographs	7, 8A, 8B, 10, 11, 20, 21, 23, 53: H. J. C. A. van Stekelenburg, Eindhoven
Photographs	9, 37: Strüwing, Copenhagen
Photographs	22, 25, 40, 47, 54, 75, 88, 90, 91, 92, 93, 94: L. C. Kalff
Photograph	24: Bundesdenkmalamt, Wien
Photograph	26: Fycer, Prague
Photographs	27, 42: Bildarchiv Foto Marburg
Photographs	28, 29: Rijksdienst voor monumentenzorg
Photograph	87: Erza Stoller Photographs
Photographs	67, 68: Rijksmuseum, Amsterdam
Drawings	28A, 28B, 29A, 29B, 38, 49, 51A, 52, 55, 56, 57, 58A, 58B, 59, 60, 61, 62, 63, 64, 65, 71, 72: L. C. Kalff
Lay-out	G. A. van 't Hof